Accelerates Academic Language Development

OXFORD ILLUSTRATED
SCIENCE
DICTIONARY

OXFORD
UNIVERSITY PRESS

OXFORD
UNIVERSITY PRESS

198 Madison Avenue
New York, NY 10016 USA

Great Clarendon Street, Oxford, OX2 6DP,
United Kingdom

Oxford University Press is a department of the University of Oxford. It furthers the University's objective of excellence in research, scholarship, and education by publishing worldwide. Oxford is a registered trade mark of Oxford University Press in the UK and in certain other countries

© Oxford University Press 2012

Library of Congress Cataloging-in-Publication Data

The Oxford illustrated science dictionary.
    p. cm.
  1. Science–Dictionaries. I. Oxford University Press.
    Q123.O94 2012
    503–dc23
        2011033012

The moral rights of the author have been asserted

First published in 2012

1 2 3 4 5 6 7 8 9 10   17 16 15 14 13 12

No unauthorized photocopying

General Manager, American ELT: Laura Pearson
Publisher: Stephanie Karras
Development Editor: Brandon Lord
Director, ADP: Susan Sanguily
Design Manager: Lisa Donovan
Cover Design: Yin Ling Wong
Electronic Production Manager: Julie Armstrong
Production Artist: Elissa Santos
Image Manager: Trisha Masterson
Image Editor: Liaht Pashayan
Senior Controller, Manufacturing: Eve Wong

ISBN: 978 0 19 407127 7

Printed in China

This book is printed on paper from certified and well-managed sources

ACKNOWLEDGEMENTS

Illustrations by: Argosy and OUP

The publishers would like to thank the following for their kind permission to reproduce photographs:

Cover photos: John Eastcott and Yva Momatiuk/National Geographic Stock (frogs); Chris Stein/Getty Images (Nautilus); Kenneth Sponsler/Shutterstock.com (telescope); ESA/Hubble Collaboration/Handout/CNP/Corbis (stars); Eric Isselée/istockphoto.com (parrot).

Pg. ii Reinhard Dirscherl/WaterFrame - Underwater Images/Photolibrary Group; pg. 2 Westend61/superstock ltd. (absorb); pg. 2 JG Photography/Oxford University Press (accelerate); pg. 3 Martyn Chillmaid/Science Photo Library (acid); pg. 3 Richard Packwood/Getty Images (acid rain); pg. 3 PIER/Getty Images (duck); pg. 4 Cyril Laubscher/Getty Images; pg. 5 Marcel Jancovic/Shutterstock.com (air resistance); pg. 5 Stephen Sharnoff/Getty Images (algae); pg. 6 Jack Goldfarb/Oxford University Press (alive); pg. 6 Peter Widmann/Age fotostock/

Photolibrary Group (allergy); pg. 7 Mikael Damkier/Shutterstock.com; pg. 8 Thomas Starke/Bongarts/Getty Images; pg. 9 Father Browne/Universal Images Group/Getty Images (WORD); pg. 9 Catmando/Shutterstock.com (ancient); pg. 9 BSIP Medical/Photolibrary Group (anesthetic); pg. 9 antos777/Shutterstock.com (animal); pg. 10 Franz Pfluegl/Fotolia; pg. 11 Bojan Fatur/istockphoto.com; pg. 14 Peter Burnett/istockphoto.com; pg. 15 Alexander Raths/Fotolia; pg. 16 INSADCO Photography/Alamy (balance weights); pg. 16 Sergey Goruppa/Alamy (balance electronic); pg. 16 Irochka/Fotolia (bacterium); pg. 17 Chris Pancewicz/Alamy; pg. 19 Eye of Science/Science Photo Library (methanosarcina); pg. 19 Peter Arnold, Inc./Alamy (mold); pg. 19 Arco Images GmbH/Alamy (tree); pg. 19 Juergen Berger/Science Photo Library (bacteria); pg. 19 Krzysztof Odziomek/Shutterstock.com (turtle); pg. 19 Peter Arnold, Inc./Alamy (fungus); pg. 20 Arcticphoto/Alamy; pg. 21 Dante Fenolio/Science Photo Library (blind); pg. 21 Peter Arnold, Inc./Alamy (blood); pg. 23 Dirk von Mallinckrodt/Mauritius/Photolibrary Group; pg. 24 Hypermania Images/Alamy; pg. 26 Buddy Mays/Alamy; pg. 27 AptTone/Shutterstock.com (diamond); pg. 27 Tyler Boyes/Fotolia (carbon); pg. 31 Frederic Cirou/Photoalto/Corbis UK Ltd (chemistry); pg. 31 Eric Nathan/Alamy (chlorophyll); pg. 33 Mark Newman/Photolibrary Group; pg. 34 Christine Osborne Pictures/Alamy (climate); pg. 34 Photodisc/Oxford University Press (cloud); pg. 34 Natalia Siverina/Shutterstock.com (coal); pg. 35 FLPA/Alamy; pg. 36 Corbis/Oxford University Press (comet); pg. 36 Stephen Hamilton/Garden Picture Library/Photolibrary Group (compost); pg. 37 LdF/istockphoto.com (computer); pg. 37 Daniel Taeger/Shutterstock.com (condensation); pg. 40 Rich Carey/Shutterstock.com; pg. 42 Mikael Hjerpe/Shutterstock.com (corrosion); pg. 42 Photodisc/Oxford University Press (crater); pg. 43 photofun/Shutterstock.com; pg. 45 Stephen Hamilton/Garden Picture Library/Photolibrary Group; pg. 46 Travelpix/Alamy; pg. 47 AptTone/Shutterstock.com; pg. 48 Stephen Miller/Alamy (diet); pg. 48 krynio/Fotolia (digestion); pg. 49 David Grossman/Alamy; pg. 53 Minden Pictures/Masterfile (ecology); pg. 53 Darryl Leniuk/Masterfile (ecosystem); pg. 54 Rubberball/Mike Kemp/Getty Images; pg. 56 Joe McDonald/CORBIS (endangered); pg. 56 Photodisc/Oxford University Press (energy); pg. 56 Corbis (engine); pg. 57 Richard Wilson/Alamy; pg. 59 Manuel Bellver/CORBIS; pg. 60 Russell Sadur/Getty Images (exercise); pg. 60 Robert Llewellyn/agefotostock (experiment); pg. 61 Philip Evans/Visuals Unlimited/Corbis; pg. 63 Ellen Isaacs/Alamy; pg. 64 Henrik Jonsson/istockphoto.com; pg. 65 Ingram Publishing/Photolibrary; pg. 66 Murat Besler/shutterstock.com; pg. 67 Wolfgang Kaehler/CORBIS; pg. 69 Ken Lucas/Getty Images (fossil); pg. 69 Markus Keller/agefotostock (icicles); pg. 71 Cerri, Lara/ZUMA Press/Corbis; pg. 72 Minerva Studio/shutterstock.com (gas); pg. 72 Walter Hodges/Getty Images (campfire); pg. 74 Gary Vogelmann/Alamy; pg. 75 Barnaby Chambers/Shutterstock.com (gear); pg. 75 Rido/Shutterstock.com (male); pg. 75 BananaStock/Oxford University Press (female); pg. 76 JKlingebiel/Shutterstock.com (lion); pg. 76 Skynavin/Shutterstock.com (tiger); pg. 76 Corbis Flirt/Alamy (generator); pg. 77 Keith Douglas/Alamy (geology); pg. 77 fusebulb/Shutterstock.com (germ); pg. 77 Chad Ehlers/Alamy (gestation); pg. 78 Paul Springett/Oxford University Press; pg. 79 Ingrid W./Shutterstock.com (wheat); pg. 79 Tish1/Shutterstock.com (corn); pg. 79 greenphile/Shutterstock.com (rice); pg. 79 AnutkaT/Shutterstock.com (glass); pg. 79 Gjermund Alsos/Shutterstock.com (graduated cylinder); pg. 80 asharkyu/Shutterstock.com (grassland); pg. 80 Vixit/Shutterstock.com (gravel); pg. 82 David Malan/Getty Images (habit); pg. 82 Tramper/Shutterstock.com (habitat); pg. 82 Erik Lam/Shutterstock.com (hair); pg. 83 Infomages/Shutterstock.com (hardness); pg. 83 F1Online/Oxford University Press (healthy); pg. 84 Photodisc/Oxford University Press (hear); pg. 84 Redphotographer/Shutterstock.com (heat); pg. 85 majeczka/Shutterstock.com; pg. 86 A & J Visage/Alamy (hibernate); pg. 86 Photodisc/Oxford University Press (horizon); pg. 87 Travelpix/Alamy (hostile); pg. 87 Image Source/Oxford University Press (human); pg. 88 Travelpix/Alamy (desert); pg. 88 Daniel Lohmer/Shutterstock.com (humidity); pg. 89 Dennis Hallinan/Alamy (hurricane); pg. 89 Photodisc/Oxford University Press (hydrogen); pg. 90 Christopher Stevenson/Photolibrary Group (ice); pg. 90 geoz/Alamy (igneous rock); pg. 90 Brand X Pictures/Oxford University Press (image); pg. 90 Dennis Kunkel/Phototake Science/Photolibrary Group (immune system); pg. Phototake Inc./Alamy (immunize); pg. 91 Ted Foxx/Alamy (inclined

continued on page 212

Contents

Acknowledgments

Our National Standards and Science Consultant

James A. Shymansky, Ph.D.

Dr. Shymansky is an E. Desmond Lee Professor of Science Education at the University of Missouri-St. Louis. He has authored or co-authored more than 25 books, chapters, and monographs, an elementary science series, and more than 80 journal articles. He is an active member of the National Association for Research in Science Teaching, for which he has served as a president and editor of its *Journal of Research for Science Teaching*.

The publisher would like to acknowledge the following individuals for their invaluable feedback during the development of this program.

Judy Dean: Austin Independent School District, Austin, TX

Dr. Sandra Stockdale: Collier County Public Schools, Naples, FL

Stefanie Kahl: Della Icenhower Intermediate School, Arlington, TX

Noreen N. Kraebel: Fox Hollow Elementary, Port Richey, FL

Molly Bostic: Heritage High School, Wake Forest, NC Iryna Khits: Hopewell High School, Charlotte, NC

Anastasia Babayan: Huntington Beach, CA

Lisa Davis, LaTisha Ford: James Coble Middle School, Arlington, TX

Tracy Thompson: Lowrey Middle School, Dearborn, MI

Tamara Lopez: McCoy Elementary, Orlando, FL

Janet E. Lasky: Montgomery County Public Schools, Rockville, MD

Misty Campos: Orange Unified School District, Orange, CA

Susan Welch: District School Board of Pasco County, Land O'Lakes, FL

Mamiko Nakata: Prince George's County Public Schools, Silver Spring, MD

Ranada Young: Roberts High School, Salem, OR

Anne Hagerman Wilcox, Lisa Nelson: Wendell School District, Wendell, ID

Gloria Prieto: Winter Springs, FL

Each part of the entry helps you learn the term.

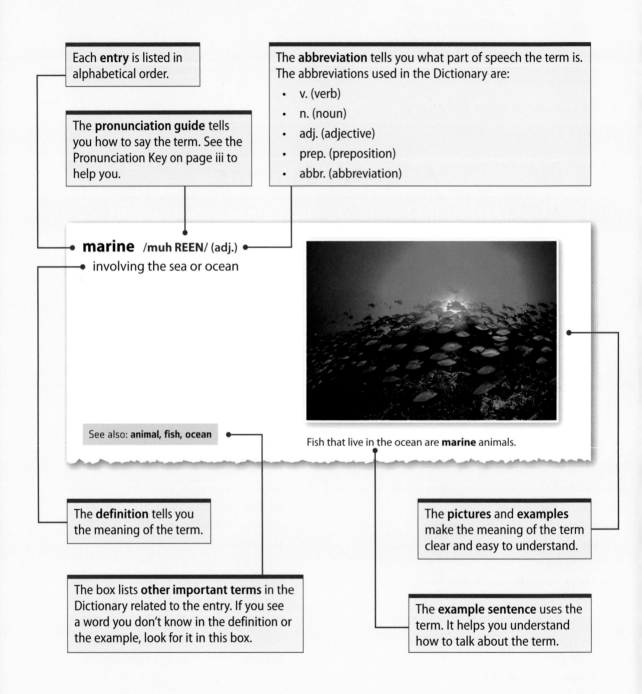

Each **entry** is listed in alphabetical order.

The **pronunciation guide** tells you how to say the term. See the Pronunciation Key on page iii to help you.

The **abbreviation** tells you what part of speech the term is. The abbreviations used in the Dictionary are:

- v. (verb)
- n. (noun)
- adj. (adjective)
- prep. (preposition)
- abbr. (abbreviation)

marine /muh REEN/ (adj.)

involving the sea or ocean

See also: **animal, fish, ocean**

Fish that live in the ocean are **marine** animals.

The **definition** tells you the meaning of the term.

The **pictures** and **examples** make the meaning of the term clear and easy to understand.

The box lists **other important terms** in the Dictionary related to the entry. If you see a word you don't know in the definition or the example, look for it in this box.

The **example sentence** uses the term. It helps you understand how to talk about the term.

Vowel Sounds	Like the Sound in . . .	Consonant Sounds	Like the Sound in . . .
a	bat, map	b	boy, job
ay	ate, say	ch	chair, lunch
ah	father, calm	d	day, mud
air	care, fair	f	fall, brief
ar	car, far	g	gone, bug
e	met, step	h	hear, hail
ee	me, equal	j	jaw, enjoy, gel
ur	fern, stir, burn	k	key, cold, took, track
i	if, fit	l	lake, tool
ī	ice, time, fly	m	my, jam
o	stop, clock	n	night, run
oh	ocean, load	ng	song, bring
or	orange, orbit	p	pay, stop
aw	jaw, talk	r	rake, press
oi	soil, boy	s	slow, bus
ow	out, flower	sh	short, bush
uh	cut, summer	t	tip, out
u	full, put	th	thick, bath
oo	soon, prove	TH	there, weather
		v	voice, save
		w	won, winter
		y	yes, young
		z	zoo, freeze
		zh	treasure

Stressed syllables appear in capital letters: /suh LOO shuhn/

abdomen /AB duh muhn/ (n.)
the part of an animal's body where the stomach is found

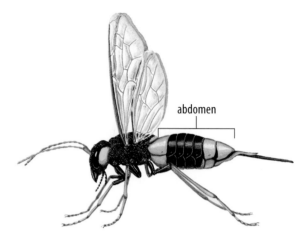

abdomen

An insect's **abdomen** is at the end of its body.

See also: **animal, insect, stomach, thorax**

absorb /uhb ZORB/ (v.)
to take in or soak up

The paper towel can **absorb** water. The table top cannot.

accelerate /ak SEL uh rayt/ (v.)
to change velocity over time

Rockets **accelerate** as they go into the air.

See also: **speed, velocity**

acid /A suhd/ (n.)
a chemical with a pH less than 7

Lemon juice is an **acid** with a pH of about 2. **Acids** turn litmus paper red.

See also: **base, chemical, litmus paper, pH**

acid rain /A suhd rayn/ (n.)
rain that contains a lot of acid because of pollution

See also: **acid, pollution**

Acid rain can harm plants and animals and damage buildings.

adaptation /a dap TAY shuhn/ (n.)
a feature that helps a living thing survive in the place it lives

See also: **environment, evolution, reproduction**

Ducks have webbed feet. This **adaptation** lets ducks move well in water.

3

adult /uh DULT/ (n.)
a fully grown animal or person

young   adult

A puppy is young. A dog is an **adult**.

See also: **caterpillar, egg, larva, life cycle**

aerodynamic
/**air roh dī NA mik**/ (adj.)
able to move easily
through the air

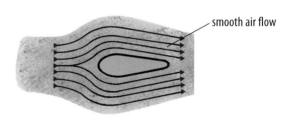

smooth air flow

An airplane wing has an **aerodynamic** shape.

age /ayj/ (n.)
how old something or someone is

The African gray
parrot can live to an
age of 100 years.

See also: **year**

air /air/ (n.)
the invisible gas that is
all around us

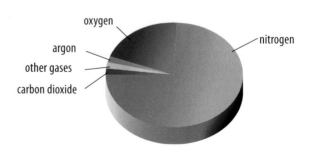

oxygen
argon
other gases
carbon dioxide
nitrogen

Air is a mixture of different gases. It is mostly nitrogen
and oxygen.

See also: **atmosphere, gas, nitrogen, oxygen**

air mass /air mas/ (n.)

a region of air above Earth that affects weather

See also: **weather**

An **air mass** can be warm or cold. It can also be wet or dry.

air pressure /air PRE shur/ (n.)

the push of the air on the area below it

lower air pressure

higher air pressure

See also: **barometer, force, pressure, weather**

Air pressure decreases as you go up a mountain.

air resistance /air ri ZIS tuhns/ (n.)

a force that slows things moving through the air

The **air resistance** on a parachute slows the fall.

See also: **aerodynamic, parachute**

algae /AL jee/ (n.)

living things that are closely related to plants

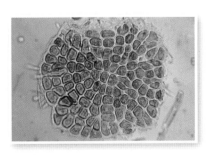

See also: **photosynthesis, plant**

Algae and plants both use photosynthesis to make sugar.

alimentary canal

/a luh MEN tuh ree kuh NAL/ (n.)

the tube from an animal's mouth to its anus

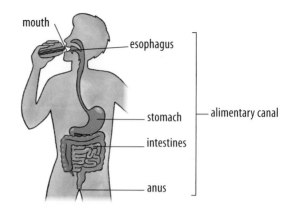

Food travels through the **alimentary canal**.

See also: **anus, digestion, intestine, mouth**

---

alive   /uh LĪV/ (adj.)

able to do things like grow, feed, and reproduce

The lizard is **alive**. The rock is not.

See also: **dead, environment, grow, reproduction**

---

allergy   /A lur jee/ (n.)

a body's reaction to a substance

Some people have an **allergy** to pollen. Allergies can be mild or serious.

See also: **pollen**

alloy /A loy/ (n.)

a mixture of two metals

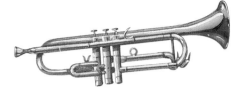

See also: **copper, metal**

Trumpets are made of brass. Brass is an **alloy** of copper and zinc.

aluminum (Al)

/uh LOO muh nuhm/ (n.)

a lightweight, nonmagnetic metal

See also: **magnetism, metal**

Cans are often made from **aluminum**.

ampere (A) /AM peer/ (n.)

a unit used to measure electric current

ampere meter

See also: **electric current, electricity, watt**

More **amperes** mean there is more electrical current.

amphibian /am FiB ee uhn/ (n.)

a kind of animal that is born in water and may live on land as an adult

back legs grow

tadpole

front legs grow

adult frog

See also: **gill, lung, mammal, metamorphosis**

A frog is an **amphibian**. It is born with gills, but it develops lungs later.

amplify /am pluh Fī/ (v.)
to make larger or louder

A megaphone works to **amplify** a person's voice.

See also: **sound waves**

amplitude /AM pluh tood/ (n.)
a measure of the height of a wave

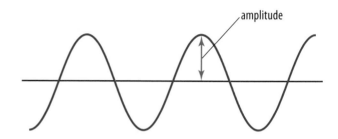

amplitude

Making a sound louder increases the **amplitude** of the sound waves.

See also: **crest, trough, wave, wavelength**

anatomy /uh NA tuh mee/ (n.)
the structure of a living thing

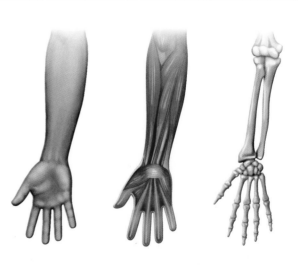

The **anatomy** of the human hand is complex.

ancestor /AN ses tur/ (n.)
a relative that lived in the past

See also: **ancient, fossil**

Family relatives have at least one **ancestor** in common.

ancient /AYN shuhnt/ (adj.)
very old or long ago

See also: **ancestor, dinosaur, fossil**

Dinosaurs lived in Earth's **ancient** past.

anesthetic /a nuhs THE tik/ (n.)
a medicine that blocks the feeling of pain

See also: **medicine**

A patient gets an **anesthetic** before an operation.

animal /A nuh muhl/ (n.)
a living thing that eats other living things

See also: **bacterium, mammal, plant, reptile**

A fish is an **animal**. There are many different kinds of **animals**.

annual /**AN yoo uhl**/ (adj.)
happening once a year

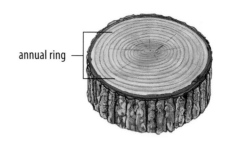

annual ring

Many trees have **annual** growth rings. They grow a new ring each year.

antennae /**an TE nee**/ (n.)
the feelers on the head
of some animals

antennae

See also: **animal, crustacean, insect**

Insects have **antennae**. Crustaceans have **antennae**, too.

anther /**AN thur**/ (n.)
the male part of a flower
that holds pollen

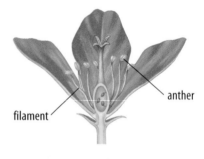

anther

filament

See also: **filament, pistil, pollen, stamen**

Bees pick up pollen from the **anther** of a flower.

antibiotic /**an tee bī AH tik**/ (n.)
a substance that kills bacteria

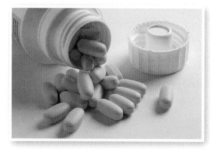

See also: **antiseptic, bacterium, germ, medicine**

Bacteria cause some illnesses. These illnesses are treated with an **antibiotic**.

antiseptic /an tuh SEP tik/ (n.)

a substance that helps prevent infection

An **antiseptic** is used on cuts and wounds. It prevents the growth of bacteria.

See also: **antibiotic, bacterium, germ, infect**

anus /AY nuhs/ (n.)

the opening at the end of the alimentary canal

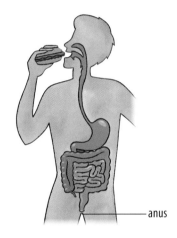

anus

See also: **alimentary canal, digest, intestine, mouth**

Solid waste leaves the body through the **anus**.

aorta /ay OR tuh/ (n.)

the artery that comes directly from the heart

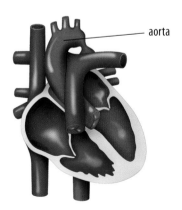

aorta

See also: **artery, blood, heart, vein**

The **aorta** is the biggest artery in the body.

aquifer /A kwuh fur/ (n.)

underground rock that is able
to hold water

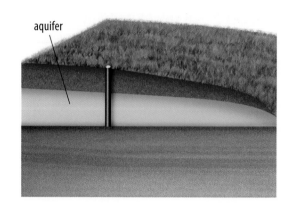

aquifer

See also: **groundwater, permeable,
surface water, watershed**

An **aquifer** contains groundwater. **Aquifers** provide
much of the drinking water in the United States.

arachnid /uh RAK nid/ (n.)

a spider or other eight-legged
animal

spider
(tarantula)

See also: **arthropod**

A spider is an **arachnid**. All **arachnids** are arthropods.

artery /AR tur ee/ (n.)

a blood vessel that carries blood
away from the heart

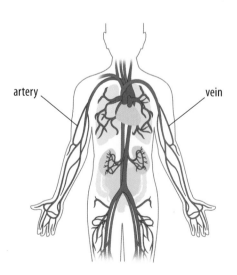

artery

vein

See also: **aorta, capillary, heart, vein**

The blood in an **artery** carries oxygen
to the body's cells.

arthropod /AR thruh pod/ (n.)

an animal with jointed legs and a hard outer skeleton

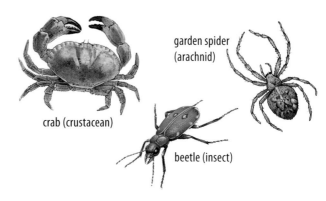

garden spider (arachnid)

crab (crustacean)

beetle (insect)

See also: **animal, arachnid, crustacean, insect**

An **arthropod** does not have a bony skeleton inside its body.

asteroid /AS tuh roid/ (n.)

a rocky or metal object that orbits the sun

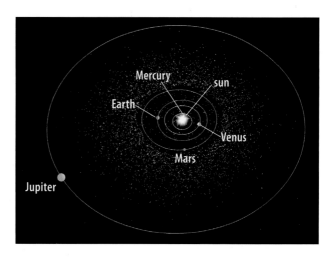

Mercury

sun

Earth

Venus

Mars

Jupiter

See also: **meteorite, orbit, planet, sun**

The area between Mars and Jupiter is called the **asteroid** belt. Most **asteroids** are there.

asthma /AZ muh/ (n.)

a condition that makes it difficult to breathe

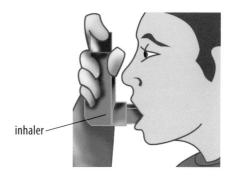

inhaler

See also: **allergy, breathe, lung**

Asthma affects the lungs. Inhalers contain medicine that can help control **asthma**.

astronomy

/uh STRAH nuh mee/ (n.)

the study of objects in space,
like planets and stars

Binoculars and telescopes are tools used
in **astronomy**.

See also: **planet, star, telescope**

atmosphere /AT muh sfeer/ (n.)

a mixture of gases that surrounds
a planet

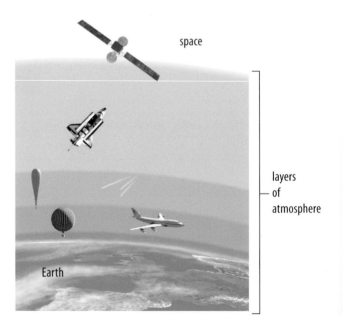

space

layers
of
atmosphere

Earth

Earth's **atmosphere** has several layers.

See also: **air, Earth, gas, planet**

atom /A tuhm/ (n.)
the smallest unit of an element

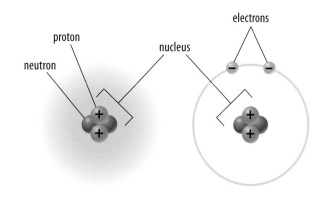

See also: **electron, element, neutron, nucleus, proton**

An **atom** contains protons, neutrons, and electrons.

automatic /aw tuh MA tik/ (adj.)
able to function on its own, without a person's control

See also: **machine**

An **automatic** dishwasher is a machine that washes dishes.

axis /AK suhs/ (n.)
an imaginary line that an object turns around

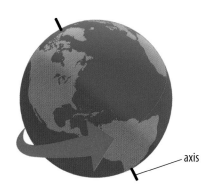

See also: **Earth, revolve, rotate, season**

Earth rotates on its **axis**.

backbone /BAK bohn/ (n.)

the set of bones in the center of the back

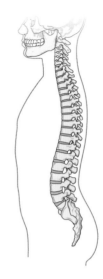

The **backbone** is not just one bone. It is made of many vertebrae.

See also: **spine, vertebra, vertebrate**

bacterium (plural bacteria)
/bak TEER ee uhm/ (n.)

a type of microorganism

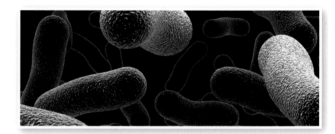

See also: **antibiotic, germ, microorganism, unicellular**

Most **bacteria** help keep the body healthy. Some can cause disease.

balance /BA luhnts/ (n.)

a tool used to measure the mass of an object

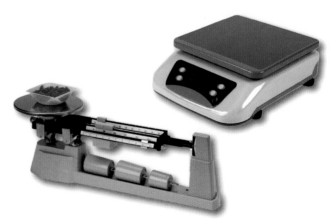

See also: **mass, microscope, ruler, weight**

One type of **balance** has weights that move. Another type of **balance** is electronic.

bark /bark/ (n.)

the outer layer of a tree

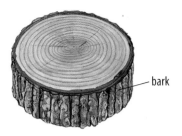

bark

See also: **tree**

Bark helps to protect a tree.

barometer /buh RAH muh tur/ (n.)

a tool used to measure
air pressure

See also: **air pressure, pressure, weather**

The **barometer** shows high pressure.
High pressure usually means clear weather.

base /bays/ (n.)

a chemical with a pH greater
than 7

See also: **acid, chemical, litmus paper, pH**

Bleach is a **base** with a pH of about 13. **Bases** turn
litmus paper green, blue, or purple.

battery /BA tuh ree/ (n.)

a device that can change
chemical energy into
electric energy

See also: **circuit, electric current,
electricity, energy**

A **battery** has a positive end and a negative end.

beak /beek/ (n.)

the mouthpart of a bird
or of some reptiles

See also: **adaptation, bird, diet, reptile**

The shape of a bird's **beak** depends on the bird's diet.

beetle /BEE tl/ (n.)

a type of insect that has hard
wing cases

See also: **animal, insect, wing**

There are hundreds of thousands of different
kinds of **beetles**.

Big Bang theory
/big bang THEER ee/ (n.)

the idea that a large explosion
may have started the universe

Many scientists think
evidence supports the
Big Bang theory.

See also: **planet, star, theory, universe**

binocular vision
/bi NAH kyuh lur VI zhuhn/ (n.)

eyesight involving two eyes
used together

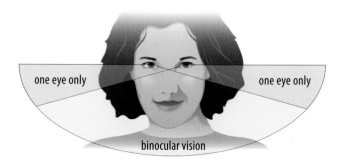

one eye only   one eye only

binocular vision

See also: **eye, senses**

Binocular vision allows a person to judge distances.

18

biodegradable
/bī oh di GRAY duh buhl/ (adj.)

able to break down
and decompose

See also: **decay, decompose**

biodegradable   not biodegradable

Plant and animal parts are **biodegradable**.
Metals, glass, and plastic are not.

biodiversity
/bī oh duh VUR suh tee/ (n.)

the variety of living things
in a particular area

See also: **rain forest, species**

There is a lot of **biodiversity** in rainforests.

biology   /bī AH luh jee/ (n.)
the study of living things

See also: **botany, zoology**

People study **biology** in every part of the world.

19

biome /BĪ ohm/ (n.)

a region with a particular climate and particular types of living things

See also: **climate**

The tundra is a **biome** found in the coldest regions of Earth.

biped /BĪ ped/ (n.)

an animal that walks on two legs

See also: **animal, bird**

A bird is a **biped**. Humans are **bipeds**, too.

bird /burd/ (n.)

an animal with two legs, wings, and feathers

See also: **animal, biped, egg, wing**

All types of **birds** lay eggs. Most **birds** can fly.

black hole /blak hohl/ (n.)

the small, dense matter
that remains after a giant
star collapses

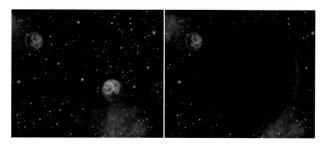

See also: **absorb, gravity, mass, star**

A **black hole** absorbs everything, including light.

bladder /BLA dur/ (n.)

the organ that stores urine

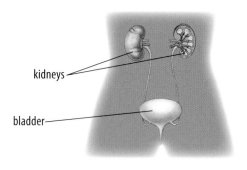

kidneys

bladder

See also: **kidney, organ**

Urine travels from the kidneys to the **bladder**.

blind /blīnd/ (adj.)

without sight

See also: **senses**

The Texas **Blind** Salamander lives in very dark caves.
It has no eyes, so it cannot see.

blood /bluhd/ (n.)

the liquid that carries oxygen,
food, and water through an
animal's body

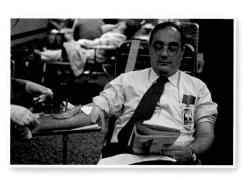

See also: **artery, capillary, oxygen, vein**

Donated **blood** helps save lives.

boiling point /BOI ling point/ (n.)

the temperature at which a liquid changes into a gas

The normal **boiling point** of water is 100°C.

See also: **freezing point, gas, liquid, melting point**

bone /bohn/ (n.)

one of the hard parts of the skeleton

Blood cells are made here.

See also: **blood, cell, skeleton**

Blood cells are made inside your **bones.**

botany /BAH tuh nee/ (n.)

the study of plants

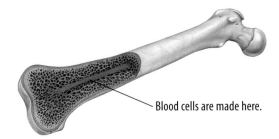

See also: **biology, zoology**

Botany is a part of biology.

brain /brayn/ (n.)

an organ made of nerve tissue that controls the body

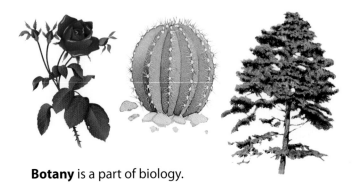

moving
speaking
touching
thinking
seeing
hearing
balance

See also: **nerve, organ, tissue**

Different parts of the **brain** do different jobs.

breathe /breeTH/ (v.)

to move air or dissolved oxygen in and out of the body

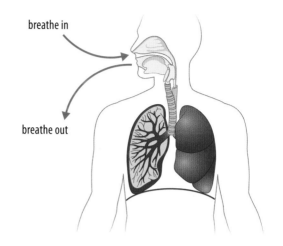

breathe in

breathe out

See also: **exhale, gill, inhale, lung**

People **breathe** through their mouths and noses.

breed /breed/ (v.)

to produce more of a certain type of plant or animal

parents   breeding   young

See also: **reproduction**

People **breed** rabbits on farms.

bud /buhd/ (n.)

an unopened flower or leaf

See also: **flower, leaf, plant**

The flower **bud** opens before the leaf **bud** does.

bulb /buhlb/ (n.)

a short underground stem

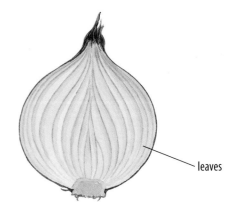

leaves

See also: **leaf, stem**

The layers of an onion **bulb** are actually thick leaves.

---

buoyancy /BOI uhnt see/ (n.)

the ability to float

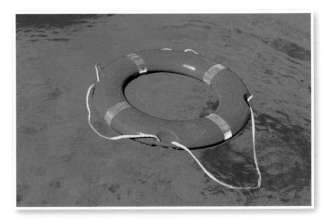

See also: **density, float**

The **buoyancy** of a life ring keeps it afloat.

---

burn /burn/ (v.)

to react with air and release energy

See also: **energy, flame, heat, light**

When candles **burn**, the flames give off light and heat.

Cc

cactus (plural cacti)
/KAK tuhs/ (n.)

a type of plant that lives in a dry environment

See also: **environment, plant**

A **cactus** usually has a thick stem to store water. There are many types of **cacti**.

caffeine /ka FEEN/ (n.)

a chemical that may make a person feel more alert

coffee   soda

See also: **chemical, drug**

Caffeine occurs naturally in coffee and tea. It is also an ingredient in some soft drinks.

calcium (Ca) /KAL see uhm/ (n.)

an element that is important to living things

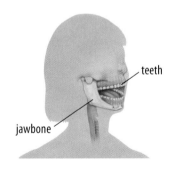

teeth

jawbone

Calcium is an important part of your bones and teeth.

See also: **element, periodic table**

Calorie (Cal) /KA luh ree/ (n.)

a unit used to measure the amount of energy in food

Nutrition Facts
Serving Size 2 crackers (14g)
Servings Per Container 21

Amount Per Serving
Calories 60   Calories from Fat 15

% Daily Value*

Total Fat 1.5g   2%

See also: **nutrient**

The label tells how many **Calories** are in the food.

25

camouflage /KA muh flahzh/ (n.)

an appearance or ability that helps an animal blend in with its surroundings

See also: **adaptation, predator, prey**

leaf insect

stick insect

Camouflage often keeps an animal safe from predators.

cancer /KAN sur/ (n.)

a disease caused by uncontrolled cell growth

See also: **disease**

Treatment for **cancer** can cause a person's hair to fall out. The hair may grow back later.

canyon /KAN yuhn/ (n.)

a landform with steep sides and a valley

See also: **valley**

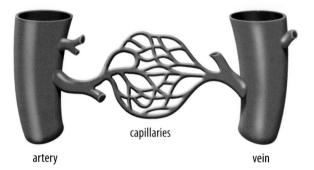

The Grand **Canyon** is a large **canyon** in Arizona.

capillary /KA puh lair ee/ (n.)

a very small blood vessel

See also: **artery, blood, vein**

capillaries

artery

vein

Materials move in and out of the blood in the **capillaries**.

carbohydrate
/kar boh HĪ drayt/ (n.)

a molecule made up of carbon, hydrogen, and oxygen

See also: **lipid, molecule, nucleic acid, protein**

Sugar, pasta, potatoes, and bread contain **carbohydrates**.

---

carbon (C) /KAR buhn/ (n.)

an element that is the main part of the molecules in living things

See also: **carbohydrate, element, molecule, periodic table, protein**

Diamond and graphite are forms of **carbon**.

---

carbon cycle
/KAR buhn SĪ kuhl/ (n.)

the movement of carbon through the air, land, water, and living things

See also: **carbon, carbon dioxide, nitrogen cycle, water cycle**

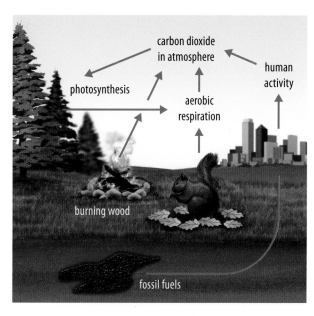

You are a part of the **carbon cycle** too.

27

carbon dioxide (CO$_2$)
/**KAR buhn dī OK sīd**/ (n.)

a gas made up of carbon and oxygen

See also: **carbon, cellular respiration, oxygen, photosynthesis**

Carbon dioxide is one of the gases in Earth's air.

---

carnivore /**KAR nuh vor**/ (n.)

an animal that eats other animals

antelope   cheetah

See also: **herbivore, omnivore**

A cheetah is a **carnivore**. It eats antelope and other animals.

---

caterpillar /**KA tur pi lur**/ (n.)

the larva of a butterfly or moth

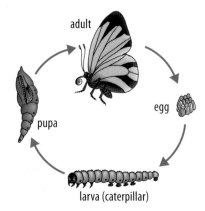

adult

egg

pupa

larva (caterpillar)

A **caterpillar** hatches from an egg. Later it turns into a butterfly or moth.

See also: **egg, larva, life cycle, pupa**

---

cell /**sel**/ (n.)

the basic unit of life

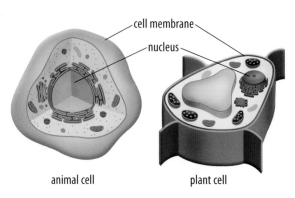

cell membrane

nucleus

animal cell   plant cell

See also: **eukaryote, nucleus, organism, prokaryote**

Some organisms are made of one **cell**. Other organisms are made of many **cells**.

cellular respiration
/SEL yuh lur res puh RAY shuhn/ (n.)

processes in a cell that break
down food and release energy

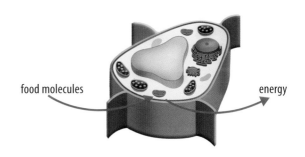

food molecules   energy

See also: **mitochondrion**

Cellular respiration happens in the mitochondria
in your cells.

Celsius (C) /SEL see uhs/ (n.)

a unit used in measuring
temperature

See also: **boiling point, Fahrenheit,
freezing point, melting point**

Water usually freezes
at 0° **Celsius** (°C).

---

center of gravity
/SEN tur uhv GRA vuh tee/ (n.)

the point at which
an object balances

high center
of gravity

low center
of gravity

See also: **gravity**

Objects with a low **center of gravity** are hard
to tip over.

---

cereal /SEER ee uhl/ (n.)

plants such as rice, oats,
wheat, and corn

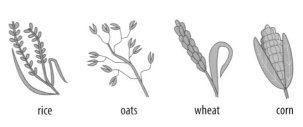

rice   oats   wheat   corn

See also: **grass**

Rice is a **cereal**. **Cereals** are very important foods.

29

cerebellum /sair uh BE luhm/ (n.)
part of the brain that helps
control movement

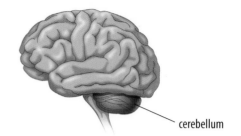

cerebellum

See also: **brain, cerebrum**

The **cerebellum** is located at the base of the brain.

cerebrum /suh REE bruhm/ (n.)
the main part of the brain

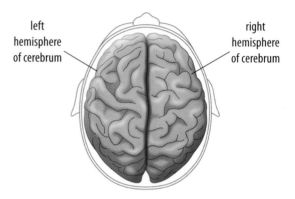

left
hemisphere
of cerebrum

right
hemisphere
of cerebrum

See also: **brain, cerebellum, hemisphere**

The **cerebrum** has a left hemisphere
and a right hemisphere.

chemical /KE mi kuhl/ (n.)
a substance with a particular
composition

radioactive   toxic   flammable

See also: **chemical change, flammable,
radioactive, toxic**

A **chemical** may be something that occurs naturally,
like water. If a **chemical** in a laboratory is dangerous,
it should have a warning label.

chemical change
/KE mi kuhl chaynj/ (n.)
a change in the molecules
that make up a substance

chemical
change

See also: **molecule, physical change**

A **chemical change** occurs when you cook an egg.

chemistry /KE mi stree/ (n.)

the study of substances and the ways they react together

You can study **chemistry** in high school.

See also: **biology, botany, geology, physics**

chlorophyll /KLOR uh fil/ (n.)

a green material in plant leaves and stems

Plants use **chlorophyll** to make sugar from the sun's energy.

See also: **chloroplast, photosynthesis, plant**

chloroplast /KLOR uh plast/ (n.)

a cell structure used for photosynthesis

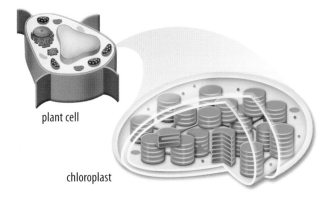

plant cell

chloroplast

See also: **algae, chlorophyll, photosynthesis, plant**

A **chloroplast** is green because it contains chlorophyll.

chromosome
/KROH muh sohm/ (n.)

a long strand of DNA

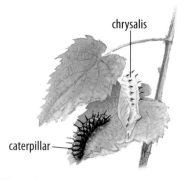

See also: **cell, DNA, gene, nucleus**

Each **chromosome** contains many different genes. Humans have 23 pairs of **chromosomes**.

chrysalis /KRI suh luhs/ (n.)

a butterfly pupa

chrysalis

caterpillar

A caterpillar becomes a butterfly inside a **chrysalis**.

See also: **caterpillar, cocoon, metamorphosis, pupa**

circuit /SUR kuht/ (n.)

a complete path that electricity can flow through

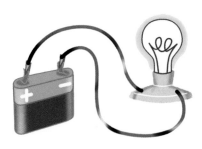

A complete **circuit** turns the light bulb on.

The light bulb stays off if the **circuit** is not complete.

See also: **battery, electricity**

circulatory system
/SUR kyuh luh tor ee SIS tuhm/ (n.)

the organ system that moves
blood throughout the body

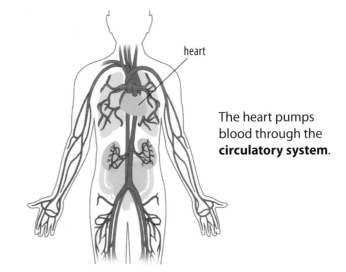

heart

The heart pumps
blood through the
circulatory system.

See also: **artery, heart, vein**

---

classify /KLA suh fī/ (v.)

to organize into categories;
to assign to a category

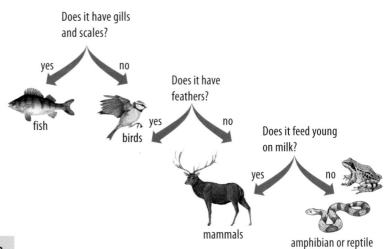

Does it have gills
and scales?

yes   no

Does it have
feathers?

fish

birds   yes   no

Does it feed young
on milk?

yes   no

mammals

amphibian or reptile

See also: **amphibian, mammal, reptile,
species**

You can **classify** an animal based on its characteristics.

---

claw /klaw/ (n.)

a sharp part on the end
of an animal's limb

An animal's **claws** may
help it to catch food.

See also: **limb, predator, prey**

33

climate /KLĪ muht/ (n.)

the usual weather in a certain place over many years

See also: **biome, desert, weather**

A desert usually has a hot, dry **climate**.

climate change
/KLĪ muht chaynj/ (n.)

a lasting change in global climate

1979   2003

Ice cover is decreasing on parts of Earth. The ice is melting because of **climate change**.

See also: **climate, weather**

cloud /klowd/ (n.)

a grouping of millions and millions of tiny water drops in the sky

See also: **atmosphere, condensation, precipitation, water cycle**

Today there are many **clouds** in the sky.

coal /kohl/ (n.)

a type of sedimentary rock that burns easily

Coal is formed from plants that lived millions of years ago.

See also: **fossil fuel, sedimentary rock**

cocoon /kuh KOON/ (n.)
a case that protects a moth pupa

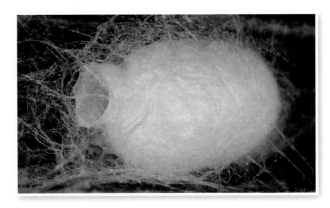

See also: **chrysalis, metamorphosis, pupa**

The moth caterpillar makes the silk that forms the **cocoon**.

cold-blooded
/**KOLD BLUH duhd**/ (adj.)
having a body temperature that is not controlled from within

See also: **warm-blooded**

Reptiles and most fish are **cold-blooded**. Some **cold-blooded** animals warm themselves with heat from the sun.

color /KUH lur/ (n.)
a characteristic of an object caused by the reflection of light

Each **color** of the parachute reflects a different wavelength of light.

See also: **light, prism, reflect, spectrum, wavelength**

comet /KOM it/ (n.)

a ball of ice, dust, and rocky particles that orbits the sun

See also: **orbit, sun**

A very bright **comet** looks like it has a tail.

compass (magnetic)
/KUM puhs/ (n.)

a tool used to determine directions

See also: **magnet, magnetic pole, magnetism**

The needle on a **compass** points north. The needle is attracted to Earth's magnetic pole.

compost /KOM pohst/ (n.)

decomposed plant material

Compost can be used to fertilize the soil in a garden.

See also: **decompose, soil**

compound /KOM pownd/ (n.)

a combination of two or more elements

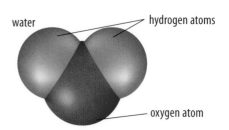

water

hydrogen atoms

oxygen atom

Water is a **compound** made up of hydrogen and oxygen.

See also: **atom, carbon dioxide, element**

computer /kuhm PYOO tur/ (n.)
an electronic device that can use and store data

See also: **data, technology**

A **computer** is a type of technology. It may be used for communication. It may also be used for graphics, data processing, and other jobs.

---

concave /kon KAYV/ (adj.)
curved inward

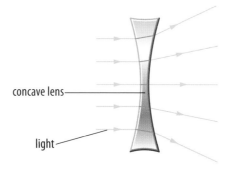

concave lens

light

See also: **convex, lens, mirror**

A **concave** lens spreads light rays.

---

concentration
/kon suhn TRAY shuhn/ (n.)
the amount of solute in a particular amount of solution

sugar

water

See also: **dilute, solute, solution, sugar**

Adding sugar increases the **concentration** of a sugar solution.

condensation
/kon den SAY shuhn/ (n.)

the change from a gas to a liquid

See also: **evaporation, gas, liquid, water vapor**

The **condensation** of water vapor causes water droplets to form.

conduct /kuhn DUHKT/ (v.)

to allow heat or electricity to pass through

Metals **conduct** electricity.

See also: **conduction, electricity, metal**

conduction
/kuhn DUHK shuhn/ (n.)

the transfer of heat between matter

metal spoon

The liquid is hot. It heats the spoon by **conduction**.

See also: **convection, matter, radiation, solid**

conifer /KON uh fur/ (n.)

a type of plant that forms seeds within cones

cone

A Scots pine tree is one example of a **conifer**.

See also: **deciduous, flower, plant, seed**

conservation

/**kon sur VAY shuhn**/ (n.)

the care and protection of resources and living things

mountain gorilla

tiger

panda

See also: **extinct**

Conservation helps to protect these animals from extinction.

constellation

/**kon stuh LAY shuhn**/ (n.)

a group of stars that forms a pattern in the sky

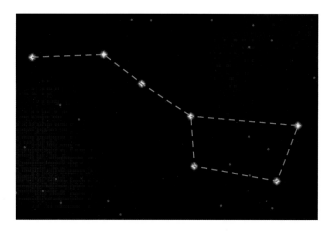

See also: **star**

The Big Dipper is a **constellation**. It contains seven stars.

consumer /**kuhn SOO mur**/ (n.)

a living thing that eats other living things

See also: **carnivore, decomposer, herbivore, producer**

A **consumer** may eat plants, animals, or both.

39

convection /kuhn VEK shuhn/ (n.)

the transfer of heat through
a liquid or gas

See also: **conduction, heat, radiation**

Heat transfer in the water happens because
of **convection**.

convex /kon VEKS/ (adj.)

curved outward

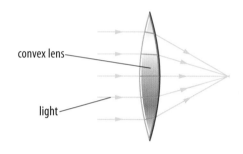

convex lens

light

See also: **concave, lens, mirror**

A **convex** lens makes light rays come together.

copper (Cu) /KOP ur/ (n.)

a metal that is easy to bend
and conducts electricity well

copper wires

See also: **conduction, electricity, metal**

Electric wires are usually made of **copper**.

coral /KOR uhl/ (n.)

a structure found in tropical seas
made up of tiny animals

See also: **animal, ecosystem, ocean**

A **coral** is made up of many tiny animals living together.

core /cor/ (n.)

the center part of something
that is round

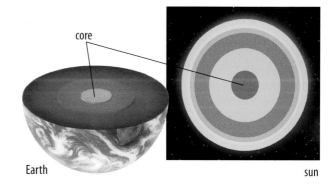

core

Earth

sun

See also: **Earth, fusion, sun**

Earth's **core** is hot. The hottest part of the sun
is the **core**.

---

cornea /COR nee uh/ (n.)

the clear outer surface of the eye
that covers the iris and pupil

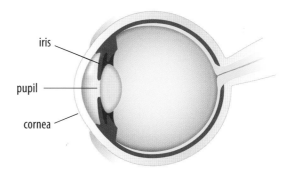

iris

pupil

cornea

See also: **iris, pupil**

The **cornea** helps focus light that enters the eye.

---

corona /kuh ROH nuh/ (n.)

the outermost layer
of the sun's atmosphere

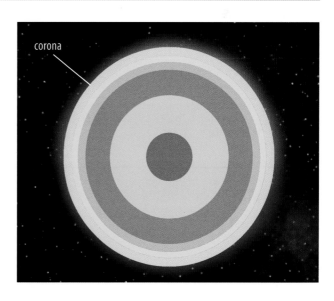

corona

See also: **core, sun**

The temperature of the **corona** is very high. It is
higher than the temperature of the sun's surface.

corrosion /kuh ROH zhuhn/ (n.)
a chemical change that breaks
down metal

See also: **chemical change, metal**

Rust is an example of **corrosion**.

cranium /KRAY nee uhm/ (n.)
the bones of the skull that
protect the brain

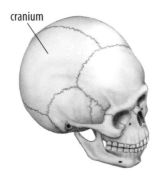

cranium

The **cranium** makes up
the top part of the skull.

See also: **skull**

crater /KRAY tur/ (n.)
a landform created by a large
object hitting Earth

See also: **meteorite**

There is a large **crater** near Flagstaff, Arizona. It was
caused by a meteorite hitting the ground.

crest /krest/ (n.)
the highest point of a wave

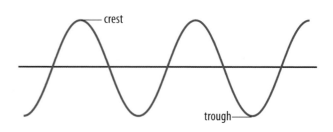

crest

trough

See also: **amplitude, trough, wave,
wavelength**

A wave's **crest** is opposite the wave's trough.

crust /cruhst/ (n.)

the outermost layer of Earth

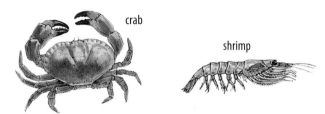

crust   mantle   core

People live on
Earth's **crust**.

See also: **core, Earth, mantle**

crustacean /kruhs TAY shuhn/ (n.)

a small animal with a hard shell
and many legs

crab

shrimp

See also: **arthropod, ocean**

A **crustacean** is a type of arthropod. Most
crustaceans live in the ocean.

crystal /KRIS tuhl/ (n.)

a solid made up of atoms that
make a repeating pattern

See also: **mineral, rock**

Quartz is a **crystal** that has six sides.

cycle /SĪ kuhl/ (n.)

a repeating pattern of events

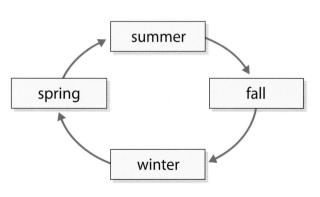

summer

spring

fall

winter

See also: **carbon cycle, life cycle,
nitrogen cycle, water cycle**

Summer is part of the **cycle** of seasons.

43

Dd

data /**DAY tuh**/ (n.)
information about a particular
situation or question

Height of Plant

	Plant A	Plant B
Day 5	5.4 cm	6.2 cm
Day 6	5.5 cm	6.5 cm
Day 7	5.5 cm	6.7 cm

Data can be recorded in many different ways.
For example, it can be words, numbers, video, or sound.

---

day /**day**/ (n.)
the time it takes for a planet
to rotate once on its axis

A **day** on Earth lasts
24 hours.

See also: **axis, night, planet, rotate**

---

dead /**ded**/ (adj.)
not able to eat, move,
or give off waste

A **dead** tree may
stay standing long
after it has died.

See also: **alive, life processes**

---

deaf /**def**/ (adj.)
without hearing

A   B   C

See also: **blind, senses**

A **deaf** person may communicate with sign language.

decay /di KAY/ (v.)

to break down through natural processes

See also: **biodegradable, decompose, decomposer**

Decomposers help make things **decay**.

---

deciduous /di SIJ oo uhs/ (adj.)

having leaves that fall off each year

See also: **evergreen, leaf**

Trees that lose their leaves each year are **deciduous**.

---

decompose /dee kuhm POHZ/ (v.)

to break down through natural processes

See also: **biodegradable, decay, decomposer**

Dead plants and animals will **decompose** over time.

---

decomposer
/dee kuhm POH zur/ (n.)

a living organism that helps to break down dead organisms

See also: **bacterium, decompose, fungus**

A worm is a **decomposer**. Bacteria and fungi are also **decomposers**.

45

delta /DEL tuh/ (n.)

a triangle of land created where
a river flows into an ocean

See also: **ocean, sediment, stream**

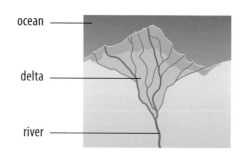

ocean

delta

river

A **delta** forms from sediment that is carried by a river.

density /DEN suh tee/ (n.)

the amount of mass within a
particular amount of space

See also: **buoyancy, mass, volume**

ice

water

metal
screws

The **density** of ice is less than the **density** of water.
This causes ice to float in water.

desert /DE zurt/ (n.)

a region that receives very little
rain each year

See also: **biome**

The Sahara is a large **desert** in northern Africa.

dew /doo/ (n.)

water drops that form because
of condensation

See also: **condensation, water, water vapor**

Dew covers plants on cool mornings.

diabetes   /dī uh BEE teez/ (n.)

a disease that affects the body's use of sugar

See also: **blood, disease, sugar**

A person with **diabetes** can use a blood test. The test reports sugar levels in the blood.

diamond   /DĪ muhnd/ (n.)

a hard, clear crystal made of carbon

Diamond is the hardest mineral.

See also: **carbon, crystal, mineral**

diaphragm   /DĪ uh fram/ (n.)

a muscle found beneath the lungs

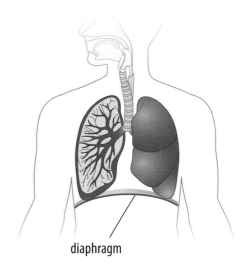

diaphragm

See also: **breathe, lung, muscle, respiratory system**

The **diaphragm** contracts to pull air into the lungs. The **diaphragm** relaxes to push air out of the lungs.

47

diet /DĪ uht/ (n.)

the food that animals or people usually eat

See also: **digestive system, nutrient, season**

The **diet** of some animals changes with the season.

diffusion /di FYOO zhuhn/ (n.)

the movement of a substance from an area with much of it to an area with little of it

See also: **concentration, molecule, osmosis**

The dye spreads through the water by **diffusion**.

digestion /dī JES chuhn/ (n.)

the process of breaking down food in the digestive system

See also: **chemical change, digestive system, physical change**

Digestion involves physical changes and involves chemical changes that begin in the mouth.

digestive system
/dī JES tiv SIS tuhm/ (n.)

the organ system that breaks down food

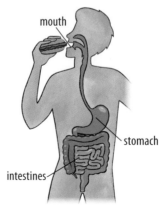

mouth

stomach

intestines

The stomach and the intestines are important parts of the **digestive system**.

See also: **digestion, intestine, stomach**

dilute /dī LOOT/ (v.)

to make less concentrated

See also: **concentration, solute, solution**

Adding more water will **dilute** a solution.

dinosaur /DĪ nuh sor/ (n.)

a type of reptile that
no longer exists

See also: **extinct, reptile**

Tyrannosaurus rex is a type of **dinosaur**. **Dinosaurs**
became extinct 65 million years ago.

disease /di ZEEZ/ (n.)

an illness or sickness

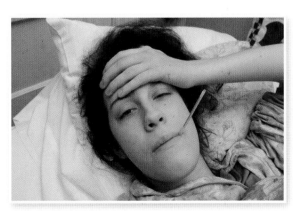

See also: **cancer, gene, heredity, parasite**

Some types of **disease** can pass from person
to person. Other **diseases** are genetic.

49

dissect /dī SEKT/ (v.)

to cut open an organism to view internal body parts

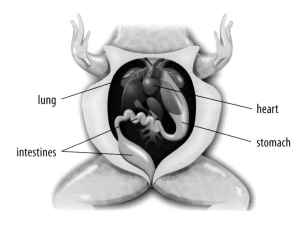

lung

heart

intestines

stomach

See also: **anatomy, heart, lung, organism**

Students sometimes **dissect** organisms in science class.

dissolve /di ZOLV/ (v.)

to become part of a liquid, forming a solution

SALT

Salt will **dissolve** in water.

See also: **salt, solute, solution**

distill /di STIL/ (v.)

to separate liquids in a liquid mixture

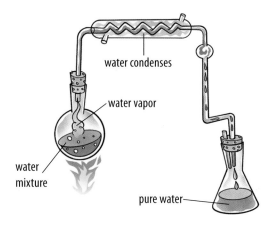

water condenses

water vapor

water mixture

pure water

See also: **condensation, evaporation, water vapor**

To **distill** water, first boil it. Then let the water vapor condense.

DNA /dee en AY/ (n.)

the genetic information that cells use to function

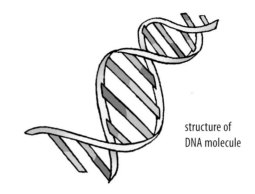

structure of
DNA molecule

See also: **chromosome, heredity, nucleic acid, nucleus**

DNA is found in the nucleus of each of your cells.
DNA stands for deoxyribonucleic acid.

---

dominant trait
/DOM uh nuhnt trayt/ (n.)

a trait that an offspring has even if only one parent has the trait

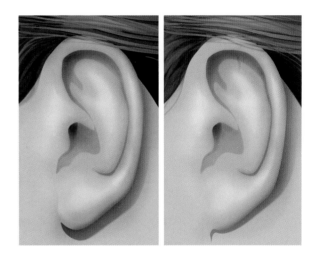

See also: **gene, heredity, recessive trait, trait**

An earlobe that hangs is a **dominant trait**.
An earlobe that is attached is a recessive trait.

---

drug /druhg/ (n.)

a substance that affects the way a person's body functions

See also: **medicine**

A **drug** is often a part of a medicine. A person can take medicine in many different ways.

51

Ee

ear /eer/ (n.)
the organ used to hear sound

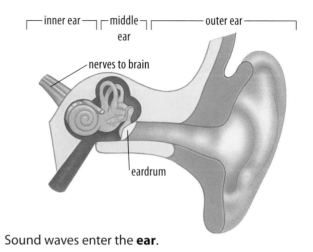

inner ear — middle ear — outer ear

nerves to brain

eardrum

See also: **deaf, organ, sound waves, vibration**

Sound waves enter the **ear**.

Earth /urth/ (n.)
the planet we live on

sun   Earth

See also: **moon, orbit, planet, sun**

Earth is the third planet from the sun.

---

earthquake /URTH kwayk/ (n.)
a sudden movement up, down, or sideways of Earth's crust

fault

See also: **crust, Earth, fault, geology**

Different types of crust movement can cause an **earthquake**.

52

echo /EK oh/ (n.)

the result of sound waves bouncing off of a surface

See also: **reflect, sound waves, wave**

You may hear an **echo** in an empty room.

eclipse /i KLIPS/ (n.)

the blocking of one object in space by another object in space

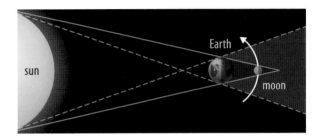

A solar **eclipse** happens when the moon is between the sun and Earth. A lunar **eclipse** happens when Earth is between the sun and the moon.

See also: **Earth, moon, sun**

ecology /ee KOL uh jee/ (n.)

the study of living things and the way they live in their environment

See also: **biology, botany, environment**

You may learn about **ecology** in science class.

ecosystem /EE koh sis tuhm/ (n.)

all the living and non-living things in a particular area

See also: **biodiversity, biome, environment, habitat**

A coral reef is an **ecosystem** with a great variety of ocean life.

53

egg /eg/ (n.)

a living structure that can
develop into an embryo

See also: **fertilization, gamete, reproduction (sexual), sperm**

A female bird lays an **egg**. A chick hatches
from the **egg**.

electric charge
/i LEK trik charj/ (n.)

a property of an atom that
depends on the number
of electrons it has

+   −

Sodium ion
Na+

Chloride ion
Cl−

See also: **electron, ion, molecule, proton**

An ion is an atom that has a positive (+) or
negative (-) **electric charge**.

electric current
/i LEK trik KUR uhnt/ (n.)

the movement of electric charges

See also: **ampere, circuit, electricity, electron**

An **electric current** occurs in a complete circuit.

electricity /i lek TRI suh tee/ (n.)

a common source of energy
in homes and businesses

See also: **energy, non-renewable energy, renewable energy**

Electricity provides energy for TVs and
other machines.

electromagnet
/i lek troh MAG nit/ (n.)

a type of magnet that needs an electric current

See also: **electric current, generator, magnet**

You can make a basic **electromagnet** at home. You can use a steel nail, a battery, and some wire.

electron   **/i LEK tron/ (n.)**

a part of an atom that has a negative charge

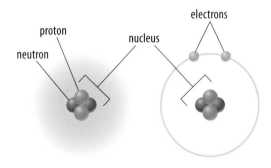

See also: **atom, neutron, nucleus, proton**

An **electron** is in the outer part of an atom.

element   **/E luh muhnt/ (n.)**

a particular type of atom

See also: **atom, carbon, molecule, periodic table**

Carbon is an **element**. There are over 100 known **elements**.

embryo   **/EM bree oh/ (n.)**

a developing plant or animal

See also: **egg, fertilization, fetus, placenta, zygote**

A human **embryo** becomes a fetus after eight weeks. This **embryo** is about six weeks old.

endangered species
/en DAYN jurd SPEE sheez/ (n.)

a species that is at high risk
of becoming extinct

See also: **extinct, species**

The mountain gorilla is an **endangered species**.
This species lives in central Africa.

energy   /E nur jee/ (n.)

the ability to make something
move or change

See also: **kinetic energy, potential energy, solar power, work**

Solar panels change **energy** from the sun
into electrical **energy**.

engine   /EN juhn/ (n.)

a device that uses energy
to move things

See also: **energy, fuel, generator, motor**

An **engine** burns fuel to function.
Some airplanes use jet **engines**.

environment
/en VĪ ruhn muhnt/ (n.)

the set of living and non-living
things that make up a region

See also: **biome, climate, ecosystem, habitat**

Soil, climate, and plants are all parts of an **environment**.

enzyme /EN zīm/ (n.)

a molecule that helps speed up
a chemical reaction

See also: **molecule, protein, reaction**

Lactase is an **enzyme**. It helps people digest cow's milk.

equator /i KWAY tur/ (n.)

an imaginary line that divides
Earth in half

equator

See also: **Earth, hemisphere, tropical**

The warmest parts of Earth are near the **equator**.
These regions are tropical.

equinox /EE kwuh noks/ (n.)

the first day of spring
and the first day of fall

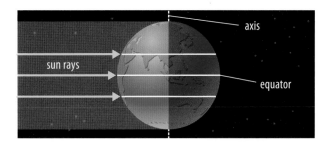

axis

sun rays

equator

See also: **axis, equator, season, solstice**

There are 12 hours of sunlight on an **equinox**.

erosion /i ROH zhuhn/ (n.)

the movement of sediment

See also: **sediment, weathering**

Water, wind, ice, and gravity can all cause **erosion**.

esophagus /i SOF uh guhs/ (n.)
the tube that connects the
mouth with the stomach

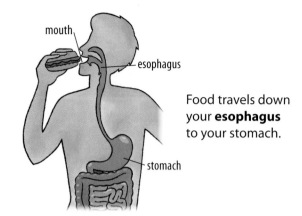

mouth

esophagus

Food travels down
your **esophagus**
to your stomach.

stomach

See also: **digestive system, mouth,
stomach**

eukaryote /yoo KAIR ee oht/ (n.)
a type of organism made of one
or more cells with a nucleus

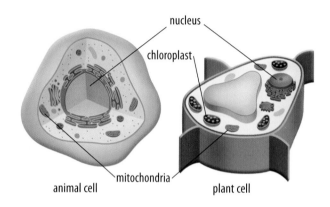

nucleus

chloroplast

animal cell

mitochondria

plant cell

See also: **chloroplast, mitochondrion,
nucleus, prokaryote**

Animals, plants, fungi, and protists are **eukaryotes.**

evaporation
/i VA puh ray shuhn/ (n.)
the change from a liquid to a gas

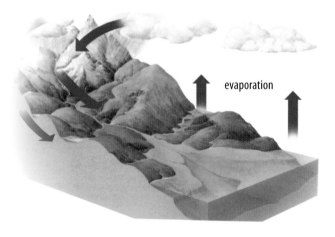

evaporation

See also: **condensation, gas, liquid,
water cycle**

Evaporation is an important process
in the water cycle.

evergreen /E vur green/ (adj.)

having needles or leaves
that stay green all year

Most conifers are
evergreen.

See also: **conifer, deciduous, plant, tree**

evidence /E vuh duhns/ (n.)

observations and data that are
used to answer a question

Evidence supports the claim that
most teenagers need more sleep.

See also: **data, hypothesis, observe**

evolution /e vuh LOO shuhn/ (n.)

the process of genetic change
over time

See also: **ancestor, fossil, natural
selection, species**

Fossils help people learn
about the **evolution**
of different organisms.

evolve /i VOLV/ (v.)

to change genetically over time

50 million years ago

35 million years ago

20 million years ago

African elephant today

See also: **adaptation, evolution,
extinct, species**

Groups of organisms **evolve** over many years.
Individual organisms do not **evolve**.

exercise /EK sur sīz/ (n.)

physical activity

Exercise helps a person stay healthy.

exhale /eks HAYL/ (v.)

to breathe out

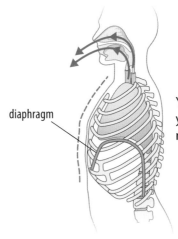

diaphragm

You **exhale** air when your diaphragm relaxes.

See also: **breathe, diaphragm, inhale, lung**

experiment /ek SPAIR uh ment/ (n.)

a test to find an explanation or answer

See also: **evidence, hypothesis, observe, theory**

Will salt water freeze as quickly as fresh water? You can do an **experiment** to find out.

explosion /ek SPLOH zhuhn/ (n.)
a sudden release of energy

An **explosion** can occur because of a chemical reaction.

See also: **flammable, pressure, volcano**

extinct /ek STINGKT/ (adj.)
no longer existing

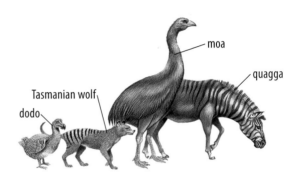

moa
quagga
Tasmanian wolf
dodo

A species is **extinct** when there are no more individuals alive.

See also: **dinosaur, evolution, fossil, fossil record**

eye /ī/ (n.)
the organ used to see

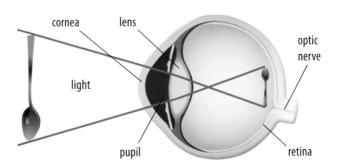

cornea
lens
optic nerve
light
pupil
retina

The **eye** senses light. The **eye** sends information to the brain through the optic nerve.

See also: **cornea, lens, pupil, retina**

Fahrenheit (F) /FAIR uhn hīt/ (n.)

a unit used to measure
temperature

See also: **boiling point, Celsius, freezing point, melting point**

Water usually freezes at 32° **Fahrenheit** (°F).

fat /fat/ (n.)

a type of tissue

walrus

See also: **cell, insulate, lipid, tissue**

Fat stores energy and insulates an animal's body.
A thick layer of **fat** helps keep a walrus warm.

fault /fawlt/ (n.)

a place where pieces
of Earth's crust meet

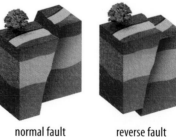

normal fault   reverse fault   strike-slip fault

See also: **crust, Earth, earthquake, geology**

An earthquake can happen along a **fault**.

feather /FE THur/ (n.)

the structure that makes up
the covering of a bird's body

Feathers help insulate a bird's body. They also help
birds to fly.

See also: **bird, insulate**

---

female /FEE mayl/ (n.)

the sex that makes eggs or seeds

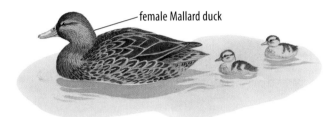

female Mallard duck

See also: **egg, male,
reproduction (sexual), sex**

An individual may be male or **female** in a species that
reproduces sexually. The **female** makes eggs or seeds.

---

fermentation
/fur men TAY shuhn/ (n.)

a process that can release energy
from sugar without oxygen

Fermentation of yeast
releases the carbon dioxide
that makes bread rise.

See also: **carbon dioxide, sugar, yeast**

---

fertilization
/fur tl uh ZAY shuhn/ (n.)

the joining of sperm and egg

fertilization

sperm + egg   zygote

See also: **egg, reproduction (sexual),
sperm, zygote**

Fertilization produces a zygote. **Fertilization** is
part of sexual reproduction.

63

fetus /FEE tuhs/ (n.)

in humans, an embryo that has
developed for more than
eight weeks

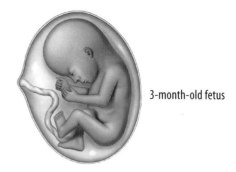

3-month-old fetus

See also: **embryo, gamete, reproduction
(sexual), zygote**

A human **fetus** is usually born after about 40 weeks.

fiber optics /FĪ bur OP tiks/ (n.)

a technology used to send
information as light signals
over long distances

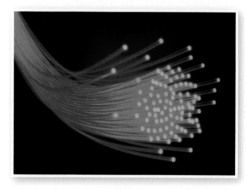

See also: **light, technology**

The Internet uses **fiber optics**. Many other
technologies use **fiber optics** too.

filament /FI luh muhnt/ (n.)

1. the part of a plant's stamen
   that supports the anther

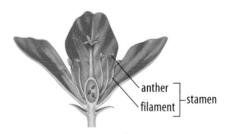

anther ⎤
filament ⎦ stamen

A **filament** is part of a flower.

2. a thin wire, as in a light bulb

filament

See also: **anther, electricity, light, stamen**

A metal **filament** in a light bulb gets very hot
and glows.

filter /FIL tur/ (n.)

a material that can separate
a solid from a liquid or gas

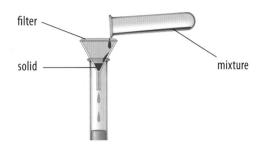

filter

solid

mixture

See also: **gas, liquid, mixture, solid, solution**

A **filter** can separate solids from a mixture.
A **filter** cannot separate solids in a solution.

fin /fin/ (n.)

a structure of fish and other
animals that live in water

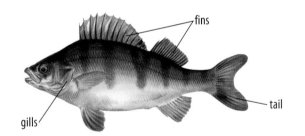

fin

See also: **fish, gill**

A fish uses its **fins** to move through water.

fish /fish/ (n.)

a kind of vertebrate
that lives in water

fins

gills

tail

See also: **fin, gill, scale, vertebrate**

A **fish** breathes through gills. Scales cover the bodies
of most **fish**.

fission /FI shuhn/ (n.)

the splitting of an atom's nucleus

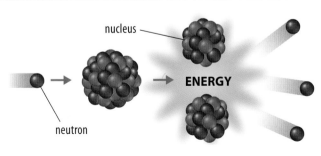

nucleus

ENERGY

neutron

See also: **atom, fusion, nuclear power,
nucleus**

Fission releases energy. Nuclear power plants use
fission to produce energy.

flame /flaym/ (n.)
a bright, visible part of a fire

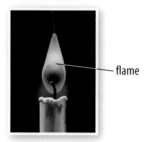

flame

See also: **chemical change, flammable, fuel**

A candle **flame** provides light.

flammable /FLA muh buhl/ (adj.)
able to burn easily

See also: **burn, flame**

A material that is very **flammable** usually has a warning label.

float /floht/ (v.)
to be supported by water or air

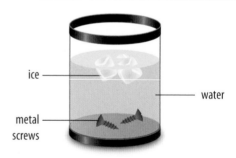

ice

water

metal screws

See also: **buoyancy, density, ice, water**

Ice is less dense than water. This makes ice cubes **float** in water.

flower /FLOW ur/ (n.)
the reproductive structure of some plants

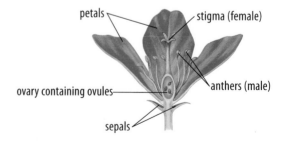

petals

stigma (female)

ovary containing ovules

anthers (male)

sepals

See also: **fertilization, fruit, pistil, stamen**

A **flower** contains a plant's reproductive structures.

fluid /FLOO id/ (n.)

a gas or liquid that moves freely

Water is a **fluid**.

See also: **gas, liquid**

---

fog /fog/ (n.)

a cloud-like mist near the ground

See also: **cloud, condensation, liquid, water**

Tiny droplets of liquid water make up **fog**.

---

food chain /food chayn/ (n.)

a way to trace energy
between organisms

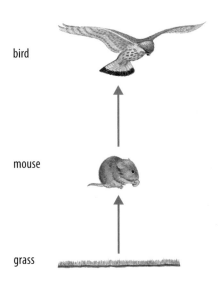

bird

mouse

grass

See also: **consumer, predator, prey, producer**

A **food chain** starts with a plant or other producers.
It then continues with consumers.

food web /food web/ (n.)

a group of food chains
that are all connected

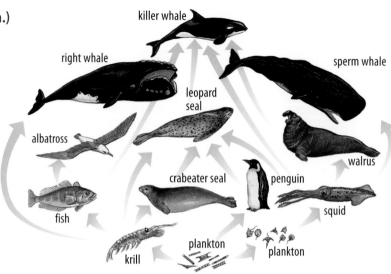

See also: **food chain, plankton**

The **food web** in the Arctic Ocean depends
on plankton.

force /fors/ (n.)

a push or a pull

See also: **friction, gravity, magnetism, simple machine**

A **force** can make an object start to move. A **force** can
also make an object slow down or change direction.

fossil /FOS uhl/ (n.)

a preserved part or trace
of a living thing

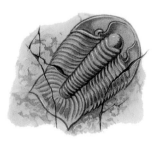

See also: **dinosaur, extinct**

This **fossil** shows a trilobite. The **fossil**
formed hundreds of millions of years ago.

fossil fuel /FOS uhl FYOO uhl/ (n.)

a fuel formed over millions
of years from the remains
of living things

Coal is a **fossil fuel**.

coal

See also: **coal, gas, non-renewable energy, oil**

fossil record
/FOS uhl REK urd/ (n.)

all known fossils and information
from fossils

The **fossil record**
suggests that birds
evolved from dinosaurs.

See also: **dinosaur, evolution, extinct, fossil**

freezing point
/FREE zing point/ (n.)

the temperature at which a liquid
changes into a solid

See also: **boiling point, liquid, melting point, solid**

The **freezing point** of water is usually 0°C.

frequency /FREE kwuhnt see/ (n.)

the number of times something happens in a given amount of time

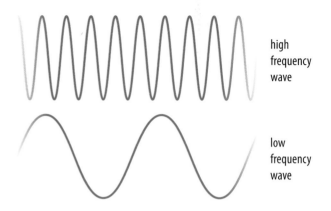

high frequency wave

low frequency wave

See also: **light, pitch, soundwaves, wave**

The pitch of a sound is the same as the **frequency** of the sound wave.

fresh water /FRESH WAW tur/ (n.)

water with very small amounts of dissolved salts

Water on Earth

fresh water
3%

salt water
97%

See also: **aquifer, groundwater, salt water, stream**

Less than 3% of Earth's water is **fresh water**. Lakes, streams, and aquifers contain **fresh water**.

friction /FRIK shuhn/ (n.)

a force that slows motion between two surfaces

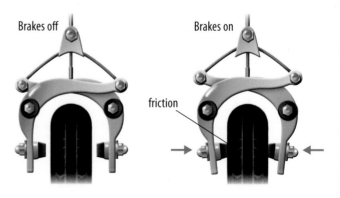

Brakes off

Brakes on

friction

See also: **force**

Bicycle brakes use **friction**. The **friction** slows the movement of the wheel.

frost /frawst/ (n.)
frozen condensation

See also: **condensation, freezing point, water vapor**

There may be **frost** on leaves on a cold morning.

fruit /froot/ (n.)
a mature plant ovary

orange

pea

strawberry

maple tree

See also: **flower, ovule, seed**

A **fruit** contains seeds. Many animals eat **fruit** for food.

71

fuel /FYOO uhl/ (n.)

a substance that releases energy

See also: **energy, fossil fuel**

Wood and gas are two kinds of **fuel**.

---

fulcrum /FUL kruhm/ (n.)

the point that a lever moves around

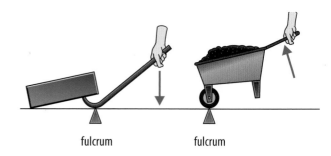

fulcrum   fulcrum

See also: **lever, simple machine**

The position of a lever's **fulcrum** defines what kind of lever it is.

---

fungus (plural **fungi**)
/FUHNG guhs/ (n.)

a type of organism that absorbs its food

See also: **absorb, decomposer, eukaryote, mushroom**

A **fungus** is a decomposer. There are many different types of **fungi**.

fuse /fyooz/ (n.)

a safety device for electrical equipment

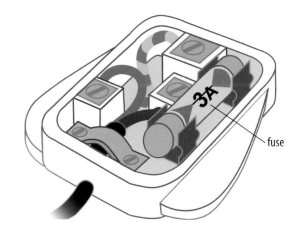

fuse

See also: **circuit, electric current, electricity**

A **fuse** contains a thin wire. The wire melts and breaks the circuit if there is too much current.

---

fusion /FYOO zhuhn/ (n.)

the joining of two or more nuclei

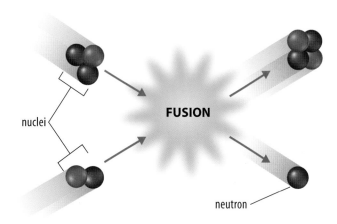

nuclei

FUSION

neutron

See also: **atom, fission, nucleus, star**

Fusion happens inside stars.

Gg

galaxy /GAL uhk see/ (n.)
a very large group of stars

See also: **star, universe**

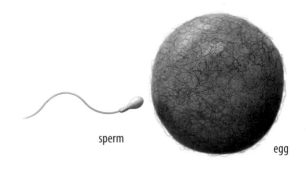

A **galaxy** may have billions of stars.

gamete /GAM eet/ (n.)
a sperm cell or egg cell

sperm

egg

See also: **egg, fertilization, meiosis, sperm**

A **gamete** is either a sperm or an egg. **Gametes** are produced by meiosis.

gas /gas/ (n.)
a form of matter that does not have a regular shape or size

The **gas** helium makes balloons float. Helium is lighter than air.

See also: **liquid, matter, solid, volume**

gear /geer/ (n.)
a wheel with raised parts
on its edge

A bicycle has **gears** to move the wheel.

gender /JEN dur/ (n.)
male or female

male · female

See also: **female, male**

A person's **gender** is male or female.

gene /jeen/ (n.)
a piece of DNA that holds
a certain trait

chromosome

DNA

gene

See also: **chromosome, DNA,
reproduction, trait**

Genes are found in the chromosomes of cells.
One kind of **gene** controls hair texture.

generator /JE nuh ray tur/ (n.)

a device that changes mechanical energy to electric energy

See also: **electricity, energy, kinetic energy**

The **generator** is an important part of a power plant.

genotype /JEE nuh tīp/ (n.)

all of the genes in an organism

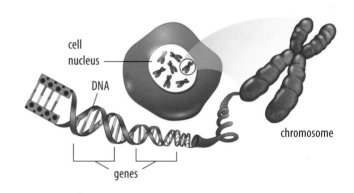

cell
nucleus

DNA

chromosome

genes

See also: **chromosome, gene, phenotype, trait**

The **genotype** includes all the genes on all the chromosomes in an organism.

genus /JEE nuhs/ (n.)

a group of similar species

lion, *Panthera leo*

tiger, *Panthera tigris*

See also: **species**

A lion and a tiger belong to the same **genus**, *Panthera*.

geology /jee AH luh jee/ (n.)

the study of Earth materials
and how they change

Geologists are people who
study **geology**. They study
rocks and minerals and
how they form.

See also: **rock cycle**

germ /jurm/ (n.)

a very small organism or a virus
that causes disease

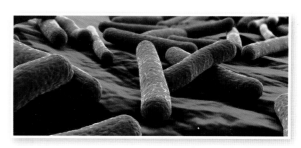

See also: **bacterium, disease,
microorganism, virus**

E. coli is a **germ** that can cause food poisoning
and make people sick.

germinate /jur muh NAYT/ (v.)

to start to grow

See also: **plant, seed**

A seed needs water and warmth to **germinate**.

gestation /je STAY shuhn/ (n.)

the time that a fetus spends
growing inside its mother's body

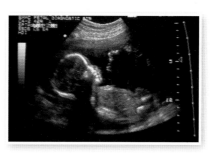

See also: **fetus, mammal, reproduction
(sexual)**

A human fetus usually spends nine months
in **gestation**.

77

gill /gil/ (n.)

an organ that allows animals
to breathe underwater

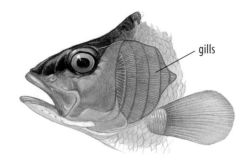
gills

See also: **fish, organ**

Fish breathe through their **gills**.

glacier /GLAY shur/ (n.)

a huge mass of ice that moves
slowly over land

See also: **ice, mountain**

A **glacier** forms over many years. **Glaciers** are
found in high mountains.

gland /gland/ (n.)

an organ that makes substances
that an organism needs to live

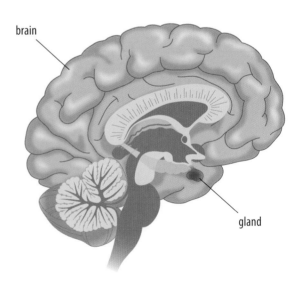
brain

gland

See also: **organ**

A **gland** near the brain controls growth.

glass /glas/ (n.)

a hard, shiny material that you can usually see through

See also: **transparent**

Glass is used to make bottles and jars.

glucose /GLOO cohs/ (n.)

a sugar that cells use for energy

See also: **carbohydrate, molecule, sugar**

Glucose is found in fruits and honey.

graduated cylinder
/GRAJ oo ay ted SI luhn dur/ (n.)

a tall container used to measure the volume of liquids

See also: **liquid, volume**

You can use a **graduated cylinder** to measure liquids.

grass /gras/ (n.)

a flowering plant with long, thin leaves

See also: **cereal, flower, leaf**

wheat   corn   rice

Wheat, corn, and rice are types of **grass**.

grassland /GRAS land/ (n.)

an area where mostly grasses grow

A **grassland** gets just enough rainfall for grasses to grow. Few trees grow in a **grassland**.

See also: **biome, grass, rainfall**

---

gravel /GRA vuhl/ (n.)

small rocks

Gravel is often found in rivers.

See also: **erosion, sediment**

---

gravity /GRA vuh tee/ (n.)

a force that pulls objects toward each other

See also: **Earth, force**

A book falls to the floor because of **gravity**.

greenhouse effect
/GREEN hows uh FEKT/ (n.)

the way Earth's atmosphere
traps heat from the sun

See also: **absorb, atmosphere,
infrared light, radiation**

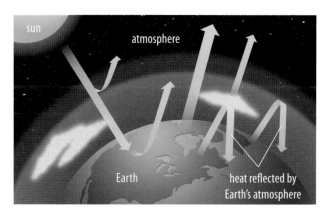

The **greenhouse effect** warms Earth. Gases in Earth's
atmosphere reflect and trap heat from the sun.

groundwater
/GROWND waw tur/ (n.)

water beneath Earth's surface

See also: **aquifer, surface water, water**

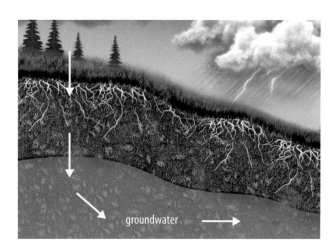

Rain moves down through the soil
to become **groundwater**.

grow /groh/ (v.)

to get larger or older

See also: **mature**

Puppies **grow** into dogs.

Hh

habit /HAB it/ (n.)
a repeated pattern of behavior

See also: **healthy**

Brushing your teeth twice a day is a good health **habit**.

habitat /HAB uh tat/ (n.)
the place where a plant
or animal lives

See also: **adaptation, environment**

A fish's **habitat** may be a lake or a stream.

hair /hair/ (n.)
special cells that grow
like threads from the skin
of mammals

See also: **cell, mammal**

Mammals have **hair** on their bodies. **Hair** helps keep them warm.

hardness /HARD nuhs/ (n.)

a measure of how
to scratch a mineral

A **hardness** scale is used to compare minerals.
A diamond is very hard. It has a **hardness** rating
of 10.

See also: **diamond, mineral**

head /hed/ (n.)

the part of the body where the
brain and the sense organs are

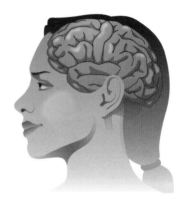

See also: **brain, organ, senses**

The human **head** contains the brain, eyes, ears,
nose, and mouth.

healthy /HEL thee/ (adj.)

when a living organism functions
the way it should

See also: **disease, exercise**

Exercising every day can help a person stay **healthy**.

hear /heer/ (v.)

to sense sound

Humans use their ears to **hear** music and other sounds.

See also: **ear, senses, sound waves**

heart /hart/ (n.)

the organ that pumps blood to all parts of the body

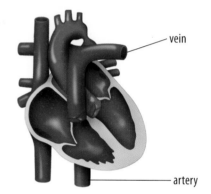

vein

artery

See also: **artery, blood, circulatory system, vein**

The **heart** pumps blood out to the body through arteries. Blood returns to the **heart** through veins.

heat /heet/ (v.)

to increase the temperature of an object or material

See also: **conduct, convection, radiation, temperature**

You can **heat** food on a stove.

heat /heet/ (n.)

a form of energy

A candle flame produces **heat** and light.

See also: **conduct, convection, energy, radiation**

hemisphere /HE muh sfeer/ (n.)

1. half of Earth

Northern Hemisphere

equator

Southern Hemisphere

North America is in the Northern **Hemisphere** of Earth.

2. half of the brain

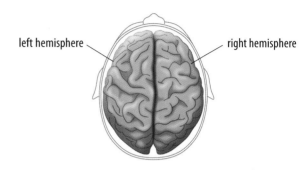

left hemisphere

right hemisphere

See also: **brain, equator**

Your brain has two **hemispheres**.

---

herbivore /HUR buh vor/ (n.)

an animal that eats only plants

See also: **carnivore, food chain, food web, omnivore**

A cow is a **herbivore**. It eats mostly grass.

---

heredity /huh RE duh tee/ (n.)

the transfer of traits from parents to offspring through genes

parent
offspring

parent

parent
offspring

offspring

See also: **DNA, gene, offspring, trait**

Offspring look like their parents because of **heredity**.

hibernate /HĪ bur nayt/ (v.)

to pass the winter in a type of deep sleep

See also: **mammal**

Some mammals **hibernate** in the winter. They are not active until spring.

HIV /h ī vee/ (n.)

the virus that causes the disease AIDS

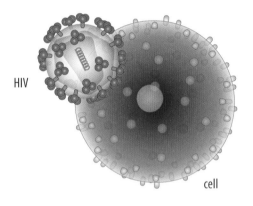

HIV

cell

See also: **cell, disease, virus**

HIV attacks body cells. **HIV** destroys the body's ability to fight diseases.

horizon /huh RĪ zuhn/ (n.)

the line where Earth seems to meet the sky

horizon

The sun appears to move toward the **horizon** at sunset.

hostile /HOS tl/ (adj.)

unfavorable for an organism

See also: **environment**

The desert is a **hostile** environment for many organisms.

human /HYOO muhn/ (n.)

a person

See also: **mammal**

A girl or boy is a **human**.

humerus /HYOO muh ruhs/ (n.)
the bone that connects the
shoulder to the elbow

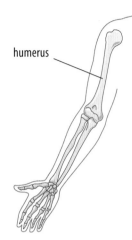

humerus

The **humerus** is the bone
in the upper arm.

See also: **bone**

humidity /hyoo MI duh tee/ (n.)
the amount of water vapor
in the air

Air in the desert has low **humidity**. Air in a rain forest
has high **humidity**.

See also: **vapor**

humus /HYOO muhs/ (n.)
the part of the soil made of
decayed plants and animals

soil with
humus

Humus makes soil
good for plants.
Humus is usually in
the top layer of soil.

rock

See also: **decay, decompose, soil**

hurricane /HUR uh kayn/ (n.)

a powerful tropical storm
with strong winds

See also: **tornado, tropical**

A **hurricane** forms over water. **Hurricane** winds
can reach more than 155 mph (249 kph).

hydrogen (H) /HĪ druh juhn/ (n.)

a gas that burns easily

Hydrogen is used as
a fuel for rockets.

See also: **burn, element, fuel, gas**

hypothesis /hī POTH uh sis/ (n.)

a possible explanation
for something based
on observations

See also: **data, evidence, experiment, theory**

A **hypothesis** can be tested in a scientific experiment.

ice /īs/ (n.)
the solid form of water

Ice forms when water freezes.

See also: **liquid, solid, water, water vapor**

igneous rock /IG nee uhs rok/ (n.)
rock that forms when melted rock cools and hardens

See also: **lava, magma, metamorphic rock, sedimentary rock**

Basalt is an **igneous rock**. It forms when lava cools.

image /I mij/ (n.)
something you see formed by reflected or refracted light

See also: **lens, light, mirror, reflect, refract**

A mirror reflects an **image** of the boy.

immune system
/i MYOON SIS tuhm/ (n.)
a group of cells and organs that helps the body fight disease

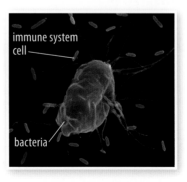

immune system cell

bacteria

The **immune system** fights bacteria and viruses.

See also: **bacterium, disease, virus**

immunize /IM yuh nīz/ (v.)

to strengthen the immune
system against a disease

See also: **disease, immune system,
influenza, vaccine**

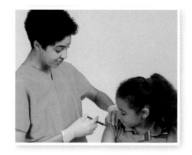

A vaccine can help
immunize a person.

impermeable

/im PUR mee uh buhl/ (adj.)

does not let water or other
fluids go through

See also: **liquid, permeable, water**

Plastic is **impermeable** to water. Paper is not.

inclined plane

/in KLĪND playn / (n.)

a simple machine shaped
like a ramp

See also: **simple machine**

An **inclined plane** makes it easier to load the truck.

incubate /IN cyuh bayt/ (v.)

to keep warm to help an
organism grow or develop

See also: **bird, egg, grow**

Birds **incubate** their eggs. They sit on the eggs
to keep them warm until the eggs hatch.

index fossil /IN deks FOS uhl/ (n.)

a fossil of an organism that was
common in the past and lived
in many places

See also: **extinct, fossil, fossil record,
geology**

An ammonite is one example of an **index fossil**.
Ammonites are extinct sea animals.

indigenous /in DI juh nuhs/ (adj.)

naturally found in a place

See also: **habitat**

Kangaroos are **indigenous** to Australia.

inertia /i NUR shuh/ (n.)

the tendency of an object to
keep moving or to stay at rest

See also: **force, matter**

Inertia keeps the book in place. It only moves when
it is pushed or pulled.

infect /in FEKT/ (v.)

to cause germs or viruses
to enter the body

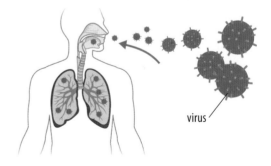

virus

See also: **bacterium, germ, influenza, virus**

The flu virus can **infect** people and make them sick.

influenza /in floo EN zuh/ (n.)

an infection of the respiratory system caused by a virus

See also: **infect, respiratory system, virus**

Influenza is also called flu. It causes a fever, sore throat, and body aches.

---

infrared light
/in fruh RED līt/ (adj.)

a type of light energy

See also: **energy, heat, light, radiation**

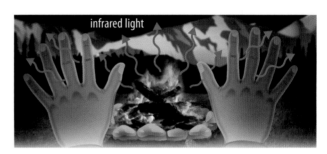

A fire produces **infrared light**. Humans cannot see **infrared light**. They feel it as heat.

---

inhale /in HAYL/ (v.)

to breathe in

See also: **diaphragm, exhale, lung, respiration**

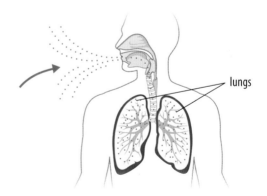

lungs

You take air into your lungs when you **inhale**.

---

insect /IN sekt/ (n.)

an animal with three body parts and six legs

See also: **abdomen, arthropod, beetle, thorax**

A bee is an **insect**.

insoluble /in SOL yuh buhl/ (adj.)
cannot be dissolved

See also: **dissolve, soluble, solute, solution**

A spoon is **insoluble** in water. It will not break down and mix with the water.

instinct /IN stinkt/ (n.)
a behavior an animal is born knowing how to do

See also: **gene, heredity, trait**

Baby sea turtles are born with an **instinct** to go to the water.

insulate /IN suh layt/ (v.)
1. to stop heat from moving from one area to another

Oven mitts **insulate**. They keep a hot pan from burning your hands.

2. to stop electricity from moving from one area to another

See also: **conduction, electricity, heat**

The plastic **insulates** the electric wire. It keeps electricity in the wire to prevent electric shocks.

intestine /in TES tuhn/ (n.)

the part of the alimentary canal between the stomach and the anus

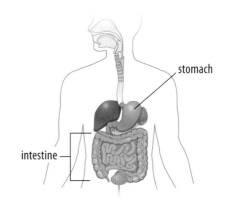

stomach

intestine

See also: **alimentary canal, digestion, stomach**

Food is digested in the **intestine**. Nutrients are also absorbed there.

invertebrate
/in VUR tuh bruht/ (n.)

an animal that does not have a backbone

See also: **backbone, vertebrate**

A worm is an **invertebrate**.

investigate /in VES tuh gayt/ (v.)

to study closely to find out something

Science students **investigate** how plants grow.

See also: **experiment, observe**

ion /ī uhn/ (n.)

an atom that has an electric charge

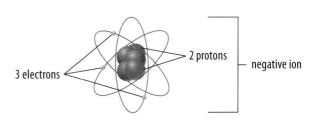

3 electrons

2 protons

negative ion

See also: **atom, electric charge, electron, proton**

Atoms usually have the same number of protons and electrons. An atom with an extra electron is a negative **ion**.

95

iris /ī ruhs/ (n.)

the colored part of the eye

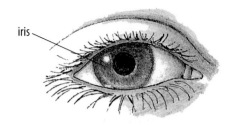

See also: **cornea, eye, pupil, retina**

The **iris** controls the amount of light entering the eye.

iron (Fe) /ī urn/ (n.)

a strong gray metal

See also: **corrosion, metal**

Iron is used to make some pots and pans. It is also used to make steel.

irreversible change
/ir i VUR suh buhl chaynj/ (n.)

a change that cannot be undone

An **irreversible change** happens when wood burns in a fire. The ashes cannot be changed back into wood.

See also: **reversible change**

Jj

jaw /jaw/ (n.)

bones in the skull that hold the teeth

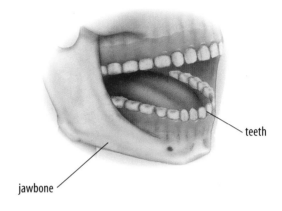

teeth

jawbone

See also: **bone, skull, tooth**

Your upper **jaw** does not move. Your lower **jaw** can move.

joint /joint/ (n.)

a place where bones meet

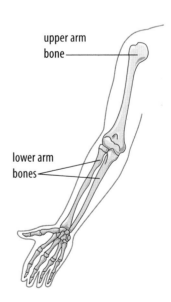

upper arm bone

lower arm bones

See also: **bone**

An elbow is a **joint**. The upper arm bone and two lower arm bones meet at the elbow.

97

kidney /KID nee/ (n.)

an organ that filters wastes
from the body

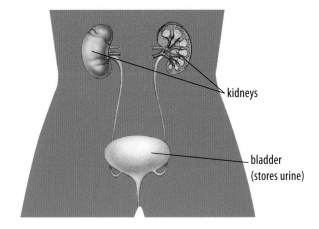

kidneys

bladder
(stores urine)

See also: **bladder**

Each **kidney** connects to the bladder.

kilogram (kg) /KI luh gram/ (n.)

a unit for measuring mass
in the metric system

1 kilogram

2.2 pounds

See also: **mass**

One thousand grams equals one **kilogram**.
One **kilogram** is about 2.2 pounds.

kilowatt (kW) /KI luh wot/ (n.)

a unit for measuring
electrical power

See also: **electricity, watt**

One thousand watts equals one **kilowatt**. A home's
electric meter measures the **kilowatts** used each hour.

kinetic energy
/kuh NE tik E nur jee/ (n.)

energy of motion

See also: **energy, potential energy**

A person riding a bicycle has **kinetic energy**.

kingdom   **/KING duhm/** (n.)

one of the six major groups used to classify living things

animal kingdom

plant kingdom

See also: **genus, species**

Animals are one major **kingdom** of living things. Plants are another.

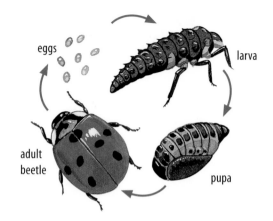

larva /LAR vuh/ (n.)

a stage in the development of some insects

See also: **beetle, insect, pupa**

A beetle **larva** develops into an adult beetle.

laser /LAY zur/ (n.)

a machine that produces a strong, narrow beam of light

See also: **light**

This device uses a **laser** to read special codes on labels.

lava /LAH vuh/ (n.)

melted rock that reaches Earth's surface

See also: **igneous rock, magma, volcano**

Lava flows from active volcanoes.

leaf　/leef/ (n.)

a part of a plant that grows out of the stem

See also: **photosynthesis, plant, stem**

A plant makes food in a **leaf** through photosynthesis.

lens　/lenz/ (n.)

1. a curved piece of clear glass or plastic that bends light

The **lens** of a magnifying glass makes an image appear larger.

2. a transparent part of the eye behind the iris

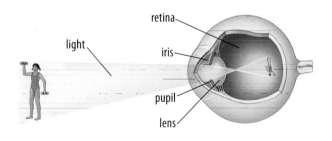

See also: **eye, image, iris, light, retina**

The **lens** focuses light rays on the retina. The light rays form an image.

lever　/LE vur/ (n.)

a simple machine made of a bar that moves around a fixed point

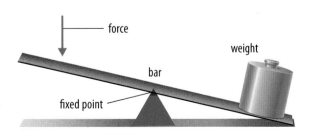

See also: **force, fulcrum, simple machine**

A **lever** can make it easier to lift or move things.

life cycle /līf SĪ kuhl/ (n.)

the stages that a living thing goes through as it grows

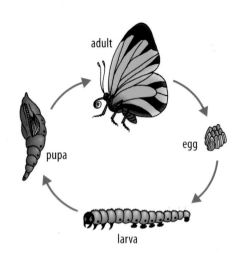

adult

egg

larva

pupa

See also: **adult, egg, larva, pupa**

A butterfly goes through four stages in its **life cycle**.

life processes
/līf PRAH se suhz/ (n.)

the activities that all living things do

See also: **grow, reproduction**

Life processes include such things as feeding, growing, and reproducing.

light /līt/ (n.)

a form of energy

See also: **energy, spectrum**

Light travels from the sun to Earth.

lightning /LĪT ning/ (n.)

a flash of light in the sky caused by a moving electrical charge

Lightning occurs during a thunderstorm.

See also: **electricity, light, thunder**

light year /līt yeer/ (n.)

the distance that light travels through space in one year

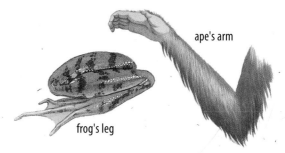

Alpha Centauri is one of the nearest stars to Earth. It is about 4.3 **light years** from the sun.

See also: **light, space**

limb /lim/ (n.)

a body part used for moving or for holding something

A leg is a **limb**. Arms are also **limbs**.

limestone /LĪM stohn/ (n.)

a type of sedimentary rock

Most **limestone** is made from the shells of ancient sea creatures. **Limestone** forms in layers.

See also: **ancient, sedimentary rock, shell**

103

limiting factor
/LI muh ting FAK tur/ (n.)

a condition that affects the survival of an organism or a population

See also: **environment, organism**

The availability of water is a **limiting factor**. Fewer animals or plants can survive when there is less water.

lipid /LI puhd/ (n.)

a fat or oil

See also: **cell, dissolve, membrane**

Most **lipids** do not dissolve in water. **Lipids** form cell structures like cell membranes.

liquid /LI kwuhd/ (n.)

a form of matter that has a definite volume but a shape that changes

See also: **gas, fluid, solid, volume**

A **liquid** flows and takes the shape of its container.

listen /LIS n/ (v.)

to make an effort to hear something

See also: **ear, hear**

You can hear quiet sounds if you **listen** closely.

litmus paper
/LIT mus PAY pur/ (n.)

a type of paper that changes color when it touches an acid or a base

See also: **acid, base, liquid**

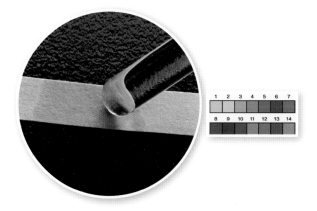

Acids make **litmus paper** turn red, orange, or yellow. Bases make **litmus paper** turn blue, green, or purple.

liver /LI vur/ (n.)

an organ that removes wastes; it also makes chemicals the body needs

See also: **organ**

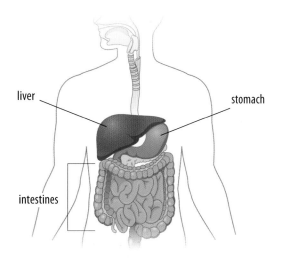

liver

stomach

intestines

The **liver** makes a chemical that helps the body digest fats.

105

look /luk/ (v.)
to make an effort
to see something

See also: **eye**

The students **look** carefully at the test tube.

lunar /LOO nur/ (adj.)
involving the moon

See also: **moon**

Scientists made a vehicle to move around
on the moon. It was called a **lunar** rover.

lung /lung/ (n.)
an organ used in breathing

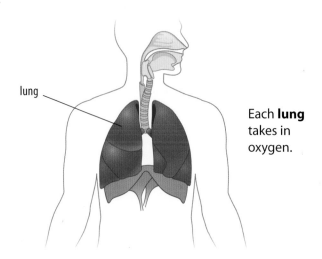

lung

Each **lung**
takes in
oxygen.

See also: **carbon dioxide, oxygen,
respiratory system**

machine /muh SHEEN/ (n.)
a device that helps do work

See also: **simple machine, wheel and axle, work**

A wheel and axle form a simple **machine**.
A doorknob is a wheel and axle.

magma /MAG muh/ (n.)
melted rock underground

See also: **igneous rock, lava, rock cycle, volcano**

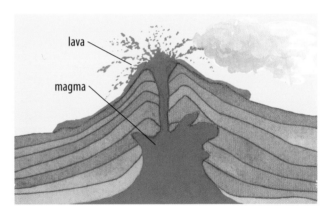

lava

magma

Magma is inside a volcano. It becomes lava
when it reaches the surface.

magnet /MAG nuht/ (n.)
an object that has
a magnetic field

See also: **iron, magnetic pole, magnetism, metal**

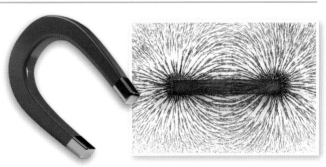

horseshoe magnet   bar magnet

A **magnet** can attract things made of iron.

107

magnetic pole
/mag NE tik pohl/ (n.)

one end of a magnet

See also: **magnet, magnetism**

A magnet has a north (N) **magnetic pole**. It also has a south (S) **magnetic pole**. A north pole and a south pole attract each other.

---

magnetism
/MAG nuh ti zuhm/ (n.)

the invisible force produced by a magnet

See also: **iron, magnet, magnetic pole**

The force of **magnetism** can attract steel to a magnet.

---

magnify   /MAG nuh fī/ (v.)

to make something appear larger

magnifying glass

A convex piece of glass will **magnify** a picture.

See also: **convex, magnifying glass**

---

magnifying glass
/MAG nuh fī ing glas/ (n.)

a convex lens that makes something look larger

See also: **convex, image, lens, magnify**

A **magnifying glass** makes the insect look larger.

male /mayl/ (n.)

the sex that produces sperm

A man or a boy is a **male**.

See also: **female, gender, sperm**

mammal /MA muhl/ (n.)

a warm-blooded animal that has hair and produces milk for its young

See also: **animal, hair, vertebrate, warm-blooded**

A dog is a **mammal**. A mother dog feeds milk to its young.

mantle /MAN tuhl/ (n.)

the layer of Earth between the crust and the core

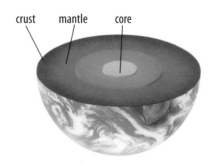

crust   mantle   core

The **mantle** contains very hot rock.

See also: **core, crust**

manual /MAN yoo uhl/ (adj.)

done by hand

See also: **automatic**

Many people collect fallen leaves with **manual** tools.

109

marine /muh REEN/ (adj.)
  involving the sea or ocean

See also: **animal, fish, ocean**

Fish that live in the ocean are **marine** animals.

mass /mas/ (n.)
  the amount of matter
  in an object

Earth   moon

See also: **inertia, kilogram, weight**

Mass is measured in grams and kilograms. **Mass** stays the same wherever the object is located.

matter /MA tur/ (n.)
  anything that has mass
  and takes up space

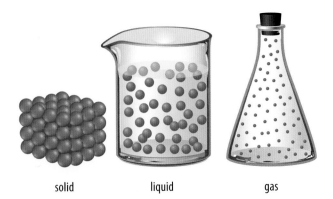

solid   liquid   gas

See also: **atom, gas, liquid, mass, solid**

Matter is made of atoms. Three states of **matter** are solid, liquid, and gas.

mature   /muh CHUR/ (adj.)
fully grown or developed

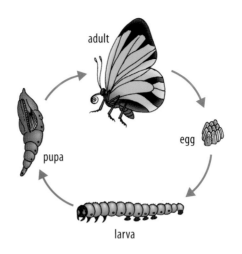

adult

egg

The adult butterfly is **mature**.

pupa

larva

See also: **adult, grow, life cycle**

maximum
/MAK suh muhm/ (adj.)
   greatest or highest possible

The daily **maximum** temperature in the Mojave Desert in the summer is about 44°C (111°F).

See also: **desert, minimum, temperature**

measure   /ME zhur/ (v.)
to find out how much
or how many

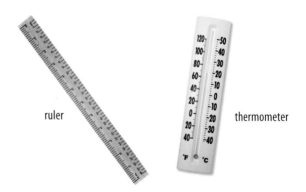

ruler

thermometer

See also: **ruler, temperature, thermometer**

A ruler is used to **measure** length. A thermometer is used to **measure** temperature.

111

medicine /ME duh suhn/ (n.)

a drug used to treat a disease
or sickness

See also: **antibiotic, disease, drug**

Aspirin is a
medicine used
to treat a headache.

---

meiosis /mī OH suhs/ (n.)

a type of cell division that
produces sperm and eggs

See also: **chromosome, egg, mitosis,
reproduction, sperm**

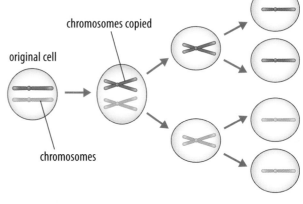

original cell

chromosomes copied

chromosomes

Meiosis is part of sexual reproduction. One original
cell produces four new cells in **meiosis**.

---

melting point
/MEL ting point/ (n.)

the temperature at which a solid
becomes a liquid

See also: **freezing point, liquid, solid,
temperature**

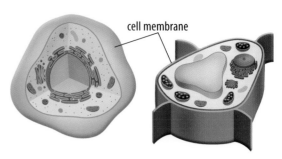

The **melting point**
for solid water (ice) is
usually 0°C (32°F).

---

membrane /MEM brayn/ (n.)

a thin layer that separates
different areas or sections

See also: **cell, permeable**

cell membrane

A cell **membrane** is a thin layer around a cell. It allows
only certain substances to pass through it.

metal /ME tuhl/ (n.)

an element that conducts heat and electricity well

See also: **alloy, conduct, electricity, heat**

A **metal** is usually shiny. Steel is a **metal** alloy.

metamorphic rock
/me tuh MOR fik rok/ (n.)

a rock formed from other rock by heat or pressure

See also: **igneous rock, rock cycle, sedimentary rock**

limestone   marble

Limestone is a sedimentary rock. Heat and pressure change it into a **metamorphic rock** called marble.

metamorphosis
/me tuh MOR fuh suhs/ (n.)

a change in the form of an animal during development

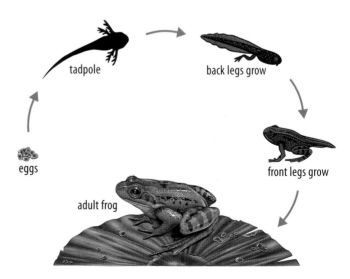

tadpole   back legs grow

eggs

front legs grow

adult frog

See also: **adult, egg, life cycle**

A tadpole changes into a frog during **metamorphosis**.

113

meteorite /MEE tee uh rīt/ (n.)

a piece of matter that reaches
Earth's surface from space

See also: **asteroid, comet, matter, space**

A **meteorite** may be made of rock or metal.
A **meteorite** is usually a part of an asteroid or comet.

microorganism
/mī kroh OR guh ni zuhm/ (n.)

an organism so small it can only
be seen with a microscope

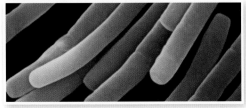

bacillus bacteria

See also: **bacterium, microscope,
organism, unicellular**

Bacteria are a type of **microorganism**.

microscope /MĪ kruh skohp/ (n.)

a tool that uses lenses to make
small things look larger

A scientist uses a
microscope to study
microorganisms.

See also: **lens, microorganism**

milk /milk/ (n.)

a white liquid made by female
mammals to feed their young

See also: **mammal**

Female mammals produce **milk**.

mineral /MI nuh ruhl/ (n.)

a natural, solid material
with a crystal structure

quartz

See also: **crystal**

Quartz is a **mineral**. Rocks are made up
of many different **minerals**.

---

minimum /MI nuh muhm/ (adj.)

least or lowest possible

See also: **maximum, temperature**

The **minimum** temperature recorded at the
South Pole is -89.2°C (-128.6°F).

---

mirror /MIR ur/ (n.)

a sheet of glass or metal
that reflects an image

A **mirror** is usually
made of glass
coated with a
silvery material.

See also: **glass, image, reflect**

---

mitochondrion
(plural **mitochondria**)
/mī tuh KAHN dree uhn/ (n.)

a structure in a cell that
makes energy

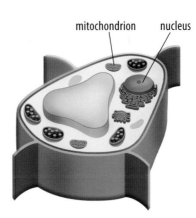

mitochondrion   nucleus

A **mitochondrion**
breaks down sugar
to make energy
for the cell.

See also: **cell, cellular respiration,
energy, nucleus**

115

mitosis /mī TOH suhs/ (n.)

a type of cell division that forms
new plant or animal cells

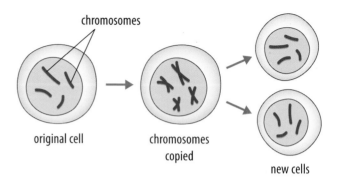

chromosomes

original cell

chromosomes
copied

new cells

See also: **cell, chromosome, meiosis**

A cell divides and forms two new cells in **mitosis**.
Each of the new cells has the same chromosomes
as the original cell.

mix /miks/ (v.)

to combine two or more
substances

SALT

Mix salt with water
to make a saltwater
solution.

See also: **compound, element, mixture,
solution**

mixture /MIKS chur/ (n.)

a combination of two
or more substances

Cookies are made
from a **mixture**.
It includes flour,
sugar, and eggs.

See also: **compound, element, mix, solution**

model /MOD uhl/ (n.)

an object built to represent
something else

A globe is a
model of Earth.

See also: **Earth**

mold /mohld/ (n.)

a type of fungus

Mold can grow
on old bread.

See also: **fungus**

mole /mohl/ (n.)

a unit used to measure
the number of molecules
in a substance

1 mole of water

See also: **atom, molecule**

One **mole** equals 6.02 x 10^{23} particles. One **mole**
of water has a mass of 18 grams.

molecule /MOL uh kyool/ (n.)

a particle made of two
or more atoms

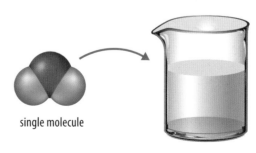

single molecule

See also: **atom, compound, hydrogen,
oxygen**

A **molecule** of water has one oxygen atom. It also has
two hydrogen atoms.

117

mollusk /MOL uhsk/ (n.)

a kind of animal with a soft body and no bones

See also: **animal, invertebrate**

A snail is a **mollusk**. Some **mollusks** have shells.

momentum /moh MEN tuhm/ (n.)

the property of a moving object; it is determined by the object's mass and velocity

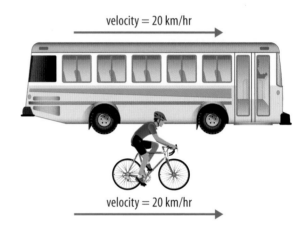

velocity = 20 km/hr

velocity = 20 km/hr

Momentum = mass × velocity

Imagine a bicycle and a bus have the same velocity. The bus has more **momentum** because it has more mass.

See also: **force, inertia, mass, velocity**

moon /moon/ (n.)

a natural satellite of a planet

See also: **lunar, orbit, reflect, satellite**

The **moon** is a satellite of Earth. It orbits Earth.

moss /mos/ (n.)

a type of small plant that does not have flowers or seeds

See also: **plant, spore**

Moss often grows in wet or shady places. **Moss** reproduces with spores.

motor /MOH tur/ (n.)

a machine that changes another form of energy into mechanical energy

See also: **electricity, energy, machine**

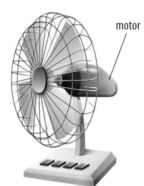

motor

A **motor** in a fan uses electricity to move the fan blades.

mountain /MOWN tuhn/ (n.)

a very large hill with steep sides

See also: **crust, plate tectonics, valley, volcano**

A volcano can form a **mountain**. Movements of Earth's crust form other **mountains**.

mouth /mowth/ (n.)

the opening an animal uses to take in food or air

See also: **air, tongue, tooth**

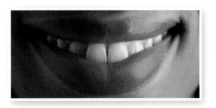

The human **mouth** includes the lips, teeth, and tongue.

119

mucus /MYOO kuhs/ (n.)

a sticky fluid that protects
some body membranes

Membranes in
the nose produce
mucus. Your nose
may produce extra
mucus when you
have a cold.

See also: **fluid, membrane**

---

multicellular
/mul ti SEL yuh lur/ (adj.)

made up of two or more cells

Moss is a **multicellular**
organism. You are
multicellular too.

See also: **moss, organism, unicellular**

---

muscle /MUH suhl/ (n.)

tissues that can move parts
of the body

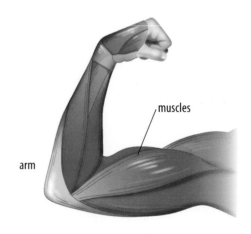

muscles

arm

See also: **bone, skeleton, tendon, tissue**

The **muscle** in your upper arm is used to lift things.

mushroom /MUHSH room/ (n.)

the reproductive structure
of some types of fungus

A **mushroom**
contains spores.
Some **mushrooms**
can be eaten, but
many **mushrooms**
are poisonous.

See also: **fungus, spore**

mutant /MYOO tuhnt/ (n.)

an organism with an unusual
change in its DNA

regular lobster   mutant blue lobster

See also: **DNA, gene, mutation, trait**

Most American lobsters are dark greenish brown.
A blue lobster is a **mutant**.

mutation /myoo TAY shuhn/ (n.)

a change in DNA; it produces
a new trait not found in
an organism's parents

regular green color   green and white color
caused by a mutation

See also: **DNA, gene, mutant, trait**

A **mutation** may change the color of a plant's leaves.

Nn

natural /NACH ur uhl/ (adj.)
produced by nature, not humans

Wood is a **natural** material. It comes from trees.

See also: **environment**

---

natural selection /NACH ur uhl suh LEK shuhn/ (n.)
the process that lets some members of a population
survive and reproduce more than others

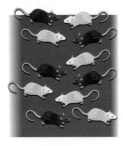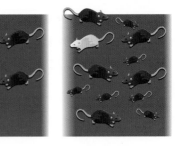

The individuals in
a population are
different from
each other.

Some individuals
survive and reproduce
more than others.

Natural selection
results in evolution.

See also: **adaptation, evolution, heredity, species**

neck /nek/ (n.)

the part of the body that joins the head to the rest of the body

neck

See also: **head, muscle**

The **neck** has muscles that make the head move.

nerve /nurv/ (n.)

a group of tissues that carries information to and from parts of the body

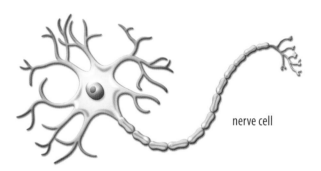

nerve cell

See also: **brain, cell, nervous system, tissue**

A **nerve** is made up of many nerve cells and other tissues. **Nerves** are found in all parts of the body.

nervous system
/NUR vuhs SIS tuhm/ (n.)

a network of nerves and organs that controls body activities

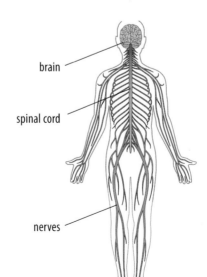

brain

spinal cord

nerves

The **nervous system** in humans includes the brain. It also includes the spinal cord and nerves.

See also: **brain, nerve, organ, spine**

net force /net fors/ (n.)

the result of two or more forces acting on an object

net force to lift:
10 lbs.

Sugar 5 lbs. Sugar 5 lbs.

Two five-pound bags of sugar require a **net force** of ten pounds to lift them.

See also: **force, newton**

---

neutron /NOO trahn/ (n.)

a particle in the nucleus of all atoms except hydrogen

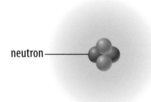

neutron —

— neutron

See also: **atom, electric charge, electron, nucleus**

A **neutron** has no electric charge.

---

newton (N) /NOO tuhn/ (n.)

a unit for measuring force

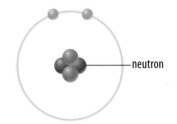

0 newtons   1 newton   2 newtons

A spring balance is used to take measurements in **newtons**.

See also: **force, measure, spring balance, weight**

---

nicotine /NI kuh teen/ (n.)

a poisonous chemical in tobacco

healthy lung   smoker's lung

See also: **chemical, drug**

Nicotine harms the lungs and other parts of the body. It makes it hard for people to stop smoking.

night /nīt/ (n.)

the time of day between sunset and sunrise

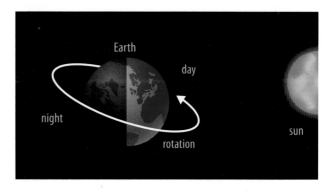

Night occurs because Earth rotates on its axis. The side of Earth that is turned away from the sun has **night**.

See also: **axis, day, rotate, sun**

nitrogen (N) /NĪ truh jen/ (n.)

a gas that makes up most of Earth's atmosphere

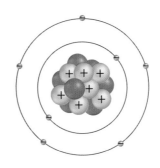

Nitrogen makes up about 80 percent of the atmosphere. The other 20 percent is mostly oxygen.

See also: **air, atmosphere, gas, oxygen**

nitrogen cycle
/NĪ truh jen SĪ kuhl/ (n.)

the process of nitrogen moving through an ecosystem

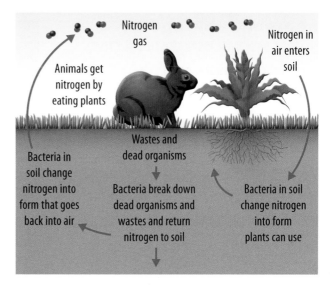

Nitrogen from the air enters the soil in the **nitrogen cycle**. Bacteria in the soil change it to a form plants can use.

See also: **bacterium, carbon cycle, ecosystem, nitrogen**

nocturnal /nok TUR nuhl/ (adj.)

active at night

Bats are **nocturnal** animals. They find food at night, and they sleep during the day.

See also: **animal, night**

nonliving /non LIV ing/ (adj.)

not able to do such things as eat, move, or give off waste

See also: **alive, environment**

Rocks are a **nonliving** part of the environment.

nonrenewable energy
/non ri NOO uh buhl E nur jee/ (n.)

cannot be replaced as fast as it is used

See also: **coal, fossil fuel, renewable energy**

Coal takes thousands of years to form. It is a source of **nonrenewable energy**.

nose /nohz/ (n.)

the body part used for breathing and smelling

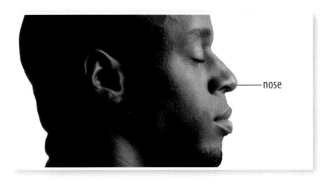

nose

See also: **breathe, lungs, respiratory system, smell**

The **nose** helps clean the air before it goes to the lungs.

nuclear power
/NOO klee ur POW ur/ (n.)

energy made by breaking apart the nucleus of atoms

See also: **atom, electricity, nucleus**

Nuclear power can be used to generate electricity.

nucleic acid
/noo KLEE ik A suhd/ (n.)

a substance important for cell activities

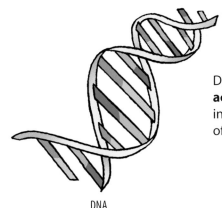

DNA is a **nucleic acid**. It is found in the nucleus of a cell.

DNA

See also: **cell, DNA, molecule, nucleus**

127

nucleus /NOO klee uhs/ (n.)

1. a large structure in a cell that controls cell activities

See also: **atom, electron, neutron, proton**

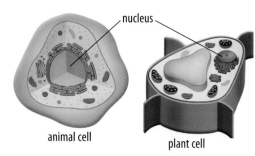

nucleus

animal cell

plant cell

The **nucleus** contains DNA.

2. the central structure in an atom

See also: **cell, DNA, gene**

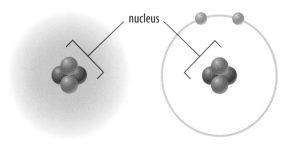

nucleus

The **nucleus** contains protons and neutrons.

---

nutrient /NOO tree uhnt/ (n.)

a substance that living things need to function and grow

See also: **vitamin**

Fruits and vegetables have **nutrients** the human body needs.

---

nylon /NĪ lon/ (n.)

a type of plastic that can be made into thread or cloth

See also: **plastic**

Nylon is a strong material. It is often used in toothbrushes and backpacks.

observe /uhb SURV/ (v.)
to look closely

You can use a microscope to **observe** a specimen.

See also: **listen, look, microscope, senses**

ocean /OH shuhn/ (n.)
a large body of salt water

See also: **marine, salt water**

The Pacific **Ocean** is the largest **ocean**.

offspring /OF spring/ (n.)
a new organism produced by reproduction

See also: **organism, reproduction, trait**

Offspring share traits with their parents.

129

oil /oil/ (n.)

a black liquid formed from
the remains of living things

See also: **coal, fossil fuel, fuel, liquid**

People pump **oil** from deep underground.

omnivore /OM nuh vor/ (n.)

an animal that eats plants
and other animals

A bear is an
omnivore. It
eats fish, fruits,
and leaves.

See also: **carnivore, herbivore**

opaque /oh PAYK/ (adj.)

does not let light through

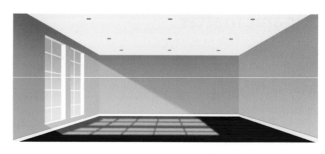

See also: **light, translucent, transparent**

A window lets light through. A wall does not.
A wall is **opaque**.

orbit /OR buht/ (n.)

the path of one object around
another object

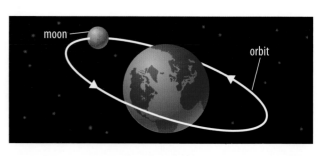

moon

orbit

See also: **Earth, moon**

The moon moves in an **orbit** around Earth.

ore /or/ (n.)

a mineral that contains a valuable
metal or other element

See also: **metal, mineral**

Copper sulfide is an **ore**. It contains the metal copper.

organ /OR guhn/ (n.)

a group of tissues that has
a specific function

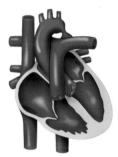

See also: **blood, heart, lung, tissue**

The heart is an **organ** that pumps blood
through the body.

organic matter
/or GA nik MA tur/ (n.)

decomposing plant and
animal material

See also: **decompose, humus, soil**

Organic matter helps make the soil healthy
for plants.

organism /OR guh ni zuhm/ (n.)

a living thing

See also: **animal, microorganism, plant**

All plants and animals are **organisms**.

131

osmosis /oz MOH sis/ (n.)
the movement of water
through a membrane

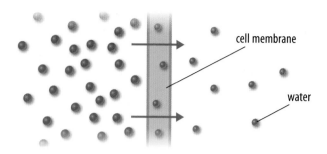

cell membrane

water

See also: **cell, liquid, membrane**

Water moves through a cell membrane by **osmosis**.

---

ossicles /AH si kuhlz/ (n.)
the small bones of the middle ear

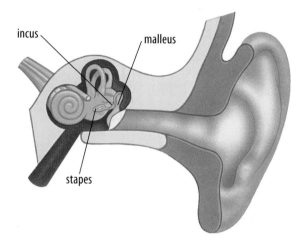

incus

malleus

stapes

See also: **ear, organ, sound, vibration**

The three **ossicles** are the malleus, incus, and stapes.

---

ovule /OH vyool/ (n.)
a plant structure that can
develop into a seed

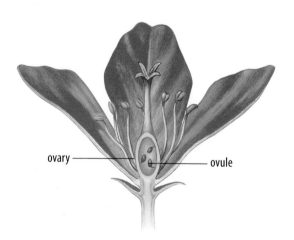

ovary

ovule

See also: **fertilization, flower, seed**

An **ovule** develops in an ovary in a flower.
After an **ovule** is fertilized it becomes a seed.

oxidation   /ok suh DAY shuhn/ (n.)

a chemical reaction in which atoms lose electrons

See also: **atom, chemical, electron, reaction**

Oxidation often involves a reaction with oxygen. Burning is an example of **oxidation**.

oxide   /OK sīd/ (n.)

a chemical compound made of oxygen and another element

See also: **chemical, compound, element, oxygen**

Rust is iron **oxide**.

oxygen (O)   /OK suh juhn/ (n.)

a gas found in Earth's atmosphere

See also: **air, atmosphere, gas, nitrogen**

Air is made up mostly of nitrogen and **oxygen**. Nitrogen makes up about 80 percent of the atmosphere. **Oxygen** makes up the rest.

ozone (O_3)   /OH zohn/ (n.)

a form of oxygen

See also: **atmosphere, gas, oxygen, ultraviolet light**

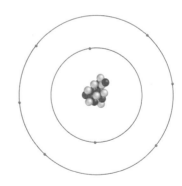

Earth   ultraviolet light

sun

ozone layer

Ozone forms naturally in the upper atmosphere. It helps block the sun's ultraviolet rays.

133

Pp

pancreas /PAN kree uhs/ (n.)
a gland near the stomach

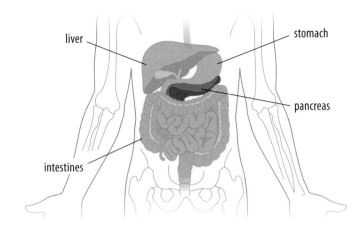

liver
stomach
pancreas
intestines

See also: **diabetes, digestion, gland**

The **pancreas** makes substances that help digest food. It also helps control blood sugar.

parachute /PAIR uh shoot/ (n.)
a device made of fabric used to slow something that is falling

See also: **gravity, momentum, velocity**

A person jumping from a plane uses a **parachute**.

parallel circuit
/PAIR uh lel SUR kuht/ (n.)
a type of electric circuit with more than one conducting path

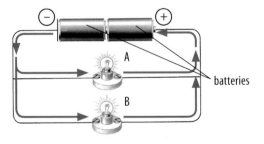

A
B
batteries

See also: **circuit, electric current, electricity, series circuit**

Most of the electric circuits in homes are **parallel circuits**.

134

parasite /PAIR uh sīt/ (n.)

an organism that lives or feeds on another organism

See also: **organism**

A tick is a **parasite**. It lives on other animals and feeds on their blood.

particle /PAR tuh kuhl/ (n.)

a very small piece of matter

See also: **microscope**

A **particle** of dust is very small. You can see it with a microscope.

peat /peet/ (n.)

a material made from partly decomposed plants

See also: **decompose, fuel, organic matter**

Peat forms in cool, wet places. It is cut from the ground and used as fuel.

pendulum /PEN juh luhm/ (n.)

a weight on the end of a string that swings back and forth

See also: **frequency, weight**

The length of the string affects how fast the **pendulum** swings.

135

periodic table
/peer ee OD ik TAY buhl/ (n.)

a chart that lists all the elements

Periodic Table of Elements

1																	18
1 H	2											13	14	15	16	17	2 He
3 Li	4 Be											5 B	6 C	7 N	8 O	9 F	10 Ne
11 Na	12 Mg	3	4	5	6	7	8	9	10	11	12	13 Al	14 Si	15 P	16 S	17 Cl	18 Ar
19 K	20 Ca	21 Sc	22 Ti	23 V	24 Cr	25 Mn	26 Fe	27 Co	28 Ni	29 Cu	30 Zn	31 Ga	32 Ga	33 As	34 Se	35 Br	36 Kr
37 Rb	38 Sr	39 Y	40 Zr	41 Nb	42 Mo	43 Tc	44 Ru	45 Rh	46 Pd	47 Ag	48 Cd	49 In	50 Sn	51 Sb	52 Te	53 I	54 Xe
55 Cs	56 Ba	57 La	72 Hf	73 Ta	74 W	75 Re	76 Os	77 Ir	78 Pt	79 Au	80 Hg	81 Tl	82 Pb	83 Bi	84 Po	85 At	86 Rn
87 Fr	88 Fa	89 Ac	104 Rf	105 Db	106 Sg	107 Bh	108 Hs	109 Mt	110 Unn	111 Rg	112 Cn	113 Uut	114 Uuq	115 Uup	116 Uuh	117 Uus	118 Uuo

Lanthanides	58 Ce	59 Pr	60 Nd	61 Pm	62 Sm	63 Eu	64 Gd	65 Tb	66 Dy	67 Ho	68 Er	69 Tm	70 Yb	71 Lu
Actinides	90 Th	91 Pa	92 U	93 Np	94 Pu	95 Am	96 Cm	97 Bk	98 Cf	99 Es	100 Fm	101 Md	102 No	103 Lr

See also: **element**

The **periodic table** lists all known elements. Groups of similar elements are grouped together.

permeable
/PUR mee uh buhl/ (adj.)

able to let liquids pass through

See also: **aquifer, impermeable, liquid, membrane**

Sponges are **permeable**. They let water pass through them.

petal   /PE tuhl/ (n.)
one of the outer parts of a flower

petal

See also: **flower, pistil, sepal, stamen**

A flower **petal** is often brightly colored.

pH /pee aych/ (n.)

a measure of the strength
of an acid or a base

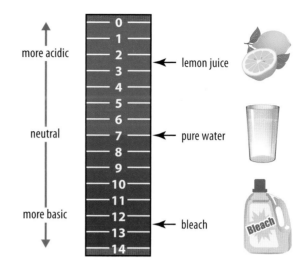

more acidic

neutral

more basic

0
1
2 ← lemon juice
3
4
5
6
7 ← pure water
8
9
10
11
12 ← bleach
13
14

See also: **acid, base, litmus paper**

A **pH** of 7 is neutral. A **pH** between 0 and 7 is an acid.
A **pH** between 7 and 14 is a base.

phase change /fayz chaynj/ (n.)

a change from one state of matter
to another

Water changes from a
liquid to a gas when it
boils. It goes through a
phase change.

See also: **gas, liquid, solid**

phenotype /FEE nuh tīp/ (n.)

the traits of an organism
that can be observed

See also: **environment, gene,
genotype, trait**

Both genes and the environment determine an
organism's **phenotype**. Eye color is a trait that is
part of a **phenotype**.

137

phloem /FLOH uhm/ (n.)

plant tissues that transport food

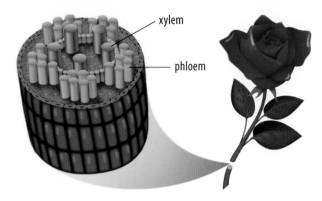

xylem

phloem

See also: **photosynthesis, plant, tissue, xylem**

Food is made in a plant's leaves. **Phloem** transports the food to other parts of the plant.

photosynthesis
/foh toh SIN thuh sis/ (n.)

the process by which plants make food from light

sunlight

leaf

sugar produced

oxygen released

carbon dioxide absorbed

water and minerals absorbed

See also: **chlorophyll, chloroplast, energy, light**

Plants use light energy to make food in **photosynthesis**. They change water and carbon dioxide into sugars and oxygen.

physical change
/FIZ uh kuhl chaynj/ (n.)

a change in the form of matter

See also: **chemical change, phase change, property**

Ice melting is a **physical change**. The chemical properties of the water do not change. The ice just changes from solid to liquid.

physics /FIZ iks/ (n.)

the study of matter and energy

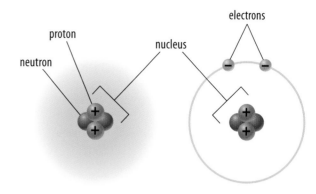

See also: **atom, energy, matter**

Physics includes the study of motion, force, heat, and light. It also includes the study of atoms.

pistil /PIS tuhl/ (n.)

the female reproductive part of a flower

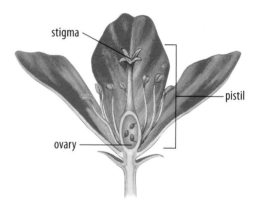

See also: **ovary, ovule, seed, stigma**

The **pistil** includes the ovary. Seeds are produced in the ovary.

pitch /pich/ (n.)

the frequency of a sound wave

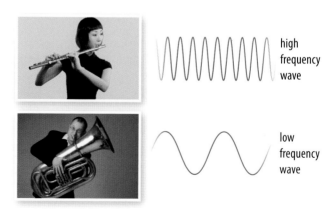

See also: **frequency, sound waves, vibration, wave**

Pitch depends on the number of vibrations per second. A high-pitched sound has many vibrations. A low-pitched sound has fewer vibrations.

placenta /pluh SEN tuh/ (n.)

an organ that develops in
the uterus during pregnancy

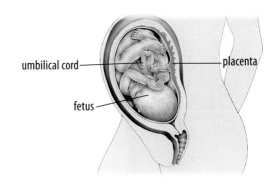

umbilical cord — placenta

fetus

See also: **fetus, mammal, umbilical
cord, uterus**

A **placenta** helps the fetus get food and oxygen
from the mother. It also helps get rid of wastes
from the fetus.

plan /plan/ (n.)

an organized way to accomplish
a goal

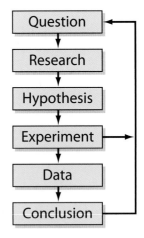

Question

Research

Hypothesis

Experiment

Data

Conclusion

See also: **experiment**

A scientist might use a list of steps to **plan**
an experiment.

planet /PLAN it/ (n.)

a large ball of rock or gas
that orbits a star

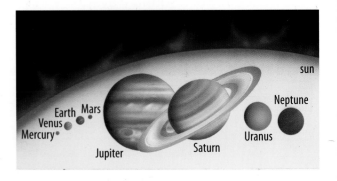

sun

Neptune

Earth Mars
Venus
Mercury

Jupiter Saturn Uranus

See also: **Earth, orbit, solar system, sun**

Earth is a **planet**. There are seven other **planets**
in our solar system.

plankton /PLANGK tuhn/ (n.)

small organisms that live in lakes, seas, and oceans

See also: **ocean, organism**

Plankton includes microscopic plants and animals. Fish and other organisms eat **plankton**.

plant /plant/ (n.)

a living thing that makes its own food from light

Trees and flowers are **plants.**

See also: **eukaryote, photosynthesis**

conifer   flowering plant

plastic /PLAS tik/ (n.)

a material made from petroleum

See also: **heat, pressure**

Heat and pressure can mold **plastic**. Bottles and food containers contain **plastic**. Some **plastic** can be recycled.

plateau /pla TOH/ (n.)

an area of high, flat land

See also: **magma**

A **plateau** may form when magma pushes the ground up.

141

plate tectonics
/**playt tek TON iks**/ (n.)

a theory that explains how pieces
of Earth's crust move

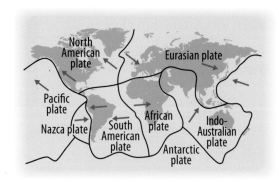

See also: **crust, earthquake, theory,
volcano**

Plate tectonics explains how Earth's crust is divided
into many pieces called plates.

pneumonia /**noo MOH nyuh**/ (n.)

an infection of the lungs

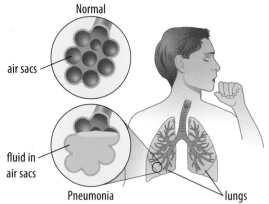

See also: **antibiotic, bacterium,
infect, lung**

Pneumonia makes air sacs in the lungs fill with fluid.
Bacteria are usually the cause of **pneumonia**.

poison /**POI zuhn**/ (n.)

a chemical that harms or kills
a living thing

See also: **chemical, toxic**

Some cleaning products contain **poison**.
They can cause sickness or death.

pollen /**POL uhn**/ (n.)

a powder produced in
the anthers of a flower

See also: **anther, fertilization, ovule, seed**

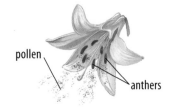

Pollen fertilizes an
ovule to make a seed.

pollinate /POL uh nayt/ (v.)

to move pollen onto the stigma of a flower

See also: **fertilization, pistil, pollen, stigma**

Bees help **pollinate** flowers.

pollution /puh LOO shuhn/ (n.)

something that harms the environment

See also: **conservation, environment, fossil fuel, gas**

Cars, trucks, and buses give off gases. The gases cause air **pollution**.

population
/pop yuh LAY shuhn/ (n.)

organisms of one species living in a habitat

See also: **habitat, organism, species**

A group of geese living in a habitat is a **population**.

potential energy
/puh TEN shuhl E nur jee/ (n.)

the energy an object has because of its position

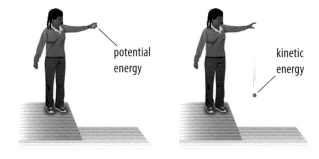

See also: **kinetic energy**

A ball raised above the ground has **potential energy**.

143

precipitation

/pri sip uh TAY shuhn/ (n.)

water that falls from clouds
to the ground

See also: **rainfall, snow**

snow

rain

Rain and snow are types of **precipitation**.

predator /PRE duh tur/ (n.)

an animal that catches and eats
other animals

See also: **carnivore, food chain,
food web, prey**

shark
(predator)

fish
(prey)

A shark is a **predator**. It catches and eats other fish.

predict /pri DIKT/ (v.)

to say what will happen

See also: **cloud, hypothesis, rainfall**

You might **predict** that it will rain if you see rain clouds.

pressure /PRE shur/ (n.)

force spread over
a particular area

The weight of an object pushes down on an area.
The weight of the object divided by the area equals
the **pressure** on the area.

See also: **force, weight**

prey /pray/ (n.)

an animal that is hunted and
eaten by another animal

barn owl
(predator)

mouse
(prey)

See also: **carnivore, food chain,
food web, predator**

A mouse is **prey** for a barn owl.

prism /PRI zuhm/ (n.)

a triangular piece of glass
that bends light waves

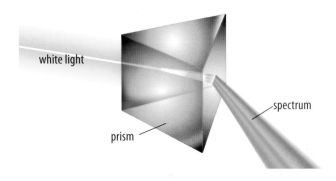

white light

spectrum

prism

See also: **light, spectrum**

A **prism** splits white light into the colors
of the spectrum.

producer /pruh DOO sur/ (n.)

an organism that uses energy
to make its own food

producer (grass)   consumer (rabbit)

See also: **consumer, decomposer,
food chain, food web**

A green plant is a **producer**.

145

prokaryote /proh KAIR ee oht/ (n.)

an organism that does not have a nucleus

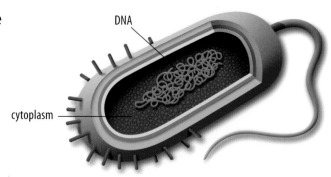

DNA

cytoplasm

See also: **bacterium, DNA, eukaryote, nucleus**

A bacterium is a **prokaryote**. Its DNA is not in a nucleus.

---

property /PROP ur tee/ (n.)

an ability or characteristic

copper wires

One **property** of copper is its ability to conduct electricity. That is why copper is used to make electrical wire.

See also: **copper, electricity**

---

protein /PROH teen/ (n.)

an important nutrient

Foods such as fish, meat, and dairy products contain **protein.**

See also: **cell, nutrient**

protist /PROH tist/ (n.)

a type of simple organism made of one or more cells

See also: **algae, eukaryote, microorganism**

A **protist** is usually a microorganism. Algae are **protists**.

proton /PROH ton/ (n.)

a particle in the nucleus of an atom

See also: **atom, electron, neutron, nucleus**

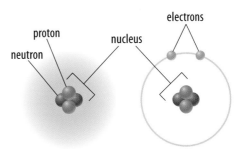

A **proton** has a positive charge.

puberty /PYOO bur tee/ (n.)

a stage of human development when sexual characteristics develop

See also: **reproduction (sexual)**

During **puberty** humans become able to reproduce. **Puberty** also causes changes in body size and shape.

pulley /PUL ee/ (n.)

a simple machine made of a wheel and a rope or chain

See also: **force, simple machine**

You can use a **pulley** to lift things.

pulse /puhls/ (n.)

a wave of pressure as blood moves through an artery

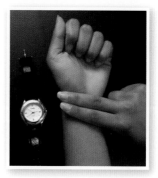

See also: **artery, blood, heart, pressure**

A heartbeat produces a **pulse**. One place to feel a **pulse** is on the inside of the wrist.

pupa /PYOO puh/ (n.)

a stage in the development of some insects

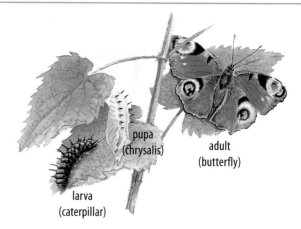

pupa
(chrysalis)

adult
(butterfly)

larva
(caterpillar)

See also: **insect, larva, life cycle, metamorphosis**

A **pupa** is the stage when a larva changes into an adult insect.

pupil /PYOO puhl/ (n.)

the hole in the center of the iris of the eye

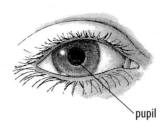

pupil

See also: **eye, iris, lens, retina**

The **pupil** lets light into the eye.

Rr

radiant /RAY dee uhnt/ (adj.)
giving off light or heat energy

See also: **energy, heat, light**

The **radiant** energy from a campfire makes you feel warm.

radiation /RAY dee ay shuhn/ (n.)
energy given off as waves
or particles

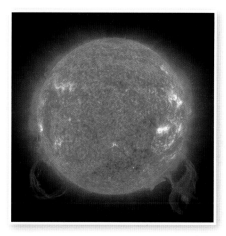

See also: **conduction, convection, energy**

Light and heat are types of **radiation**.
The sun gives off **radiation**.

radioactive
/ray dee oh AK tiv/ (adj.)
giving off energy from particles
in the nucleus of an atom

See also: **atom, element, energy, nucleus**

Some elements are **radioactive**. They give off
a dangerous form of energy.

149

rainbow /RAYN boh/ (n.)

a band of colors that appears when sunlight is refracted

See also: **color, light, refract, spectrum**

A **rainbow** forms when sunlight passes through raindrops. The sunlight splits into many colors.

---

rainfall /RAYN fawl/ (n.)

the amount of rain that falls in an area over a period of time

Average Monthly Rainfall in Suntown, USA

(bar graph: x-axis "Month" with Jan, Feb, Mar, Apr, May, Jun, Jul, Aug, Sep, Oct, Nov, Dec; y-axis "Average rainfall (inches)" from 0 to 5)

See also: **climate, precipitation, rain gauge, weather**

Scientists keep records of **rainfall**. This helps them understand weather and climate.

---

rain forest /rayn FOR uhst/ (n.)

a forest that gets over 68 inches of rainfall throughout the year

See also: **biodiversity, biome, rainfall**

A **rain forest** has tall trees that stay green all year.

---

rain gauge /rayn gayj/ (n.)

a tool for measuring rainfall

A **rain gauge** measures the amount of rainfall. A **rain gauge** can measure rainfall in inches or centimeters.

See also: **measure, rainfall, weather**

150

range /raynj/ (n.)

the difference between the smallest and largest number in a set of data

See also: **data, maximum, minimum**

| Currently **43°** | ☀ Sunny | High **55°** |
| | | Low **39°** |

The temperature **range** for today was about 16 degrees. It went from 39˚F to 55˚F.

rate /rayt/ (n.)

the change in one measurement compared to the change in a second measurement

See also: **measure, speed**

A speedometer shows the **rate** at which a car travels.

react /ree AKT/ (v.)

to change in response to something else

See also: **enzyme, reactant, reaction, stimulus**

Baking soda and vinegar **react** when mixed. The reaction produces carbon dioxide.

reactant /ree AK tuhnt/ (n.)

a substance in a chemical reaction

See also: **chemical change, corrosion, reaction**

Iron reacts with oxygen in the air to form rust. Iron is a **reactant**. Oxygen is also a **reactant**.

reaction /ree AK shuhn/ (n.)

1. a chemical change

See also: **chemical, chemical change, react, reactant**

Chemicals mix together in a chemical **reaction**. They change into new substances.

2. a response to a stimulus

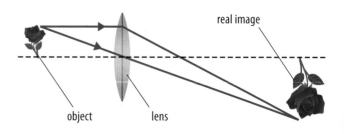

A lemon is sour. Most people have a **reaction** to the sour taste.

See also: **react, response, stimulus**

---

real image /REE uhl IM ij/ (n.)

an image formed by rays of light coming together

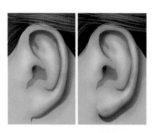

Light rays from an object go through a lens and create a **real image**. The **real image** looks like the object upside down.

See also: **image, lens, light**

---

recessive trait /ri SE siv trayt/ (n.)

a trait that an offspring will have only if both parents have the trait

See also: **dominant trait, genes, heredity, trait**

An earlobe that is attached is a **recessive trait**. An earlobe that hangs is a dominant trait.

record /ri KORD/ (v.)
to write down observations
and data

Researchers **record** data
during an experiment.

See also: **data, experiment, measure,
observe**

red blood cell /red bluhd sel/ (n.)
a cell in the blood that carries
oxygen and carbon dioxide

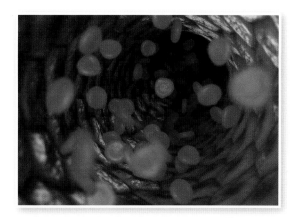

See also: **blood, carbon dioxide,
circulatory system, oxygen**

A **red blood cell** carries oxygen to cells. It carries
carbon dioxide away from cells.

reduction /ri DUHK shuhn/ (n.)
a type of chemical reaction

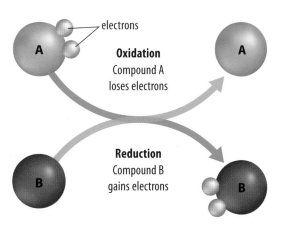

electrons

A

Oxidation
Compound A
loses electrons

A

B

Reduction
Compound B
gains electrons

B

The atoms in a compound gain electrons in **reduction**.
Reduction always happens with oxidation.

153

reef /reef/ (n.)

rock or coral near the surface
of a sea or ocean

coral

Skeletons of tiny animals
make up a coral **reef**. It is
a habitat for many animals.

See also: **ecosystem, habitat, marine, ocean**

reflect /ri FLEKT/ (v.)

to bounce back light or sound

A mirror can **reflect** light.

See also: **echo, image, mirror, refract,
sound waves**

reflex /REE fleks/ (n.)

an action that happens
automatically as a reaction
to something

Your leg moves in a **reflex** when a doctor taps below
your knee. You don't have to think about it.

See also: **reaction, stimulus**

refract /ri FRAKT/ (v.)

to bend the path of light

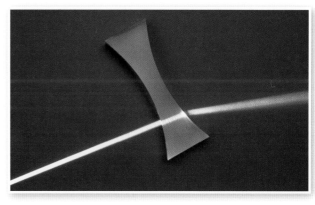

See also: **lens, light, reflect**

Light normally travels in a straight path.
A lens can **refract** light.

renewable energy
/ri NOO uh buhl EN ur jee/ (n.)

energy from resources that
cannot be used up

See also: **energy, fossil fuel,
non-renewable energy, solar power**

Sunlight and wind are sources of **renewable energy**.

reproduction (asexual)
/ree pruh DUHK shuhn (ay SEK shoo uhl)/
(n.)

a process in which one parent
organism produces offspring

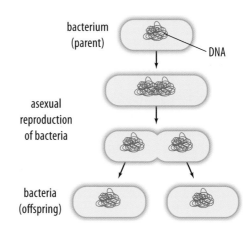

bacterium
(parent)

DNA

asexual
reproduction
of bacteria

bacteria
(offspring)

See also: **bacterium, DNA, offspring,
reproduction (sexual)**

Many single-celled organisms reproduce by
asexual reproduction.

155

reproduction (sexual)
/ree pruh DUHK shuhn (SEK shoo uhl)/ (n.)

a process in which two parent organisms produce offspring

parents   offspring

See also: **gamete, gene, meiosis, reproduction (asexual)**

Almost all animals reproduce through **sexual reproduction**. Offspring inherit their parents' genes.

---

reptile /REP tī uhl/ (n.)
a cold-blooded animal with a body covered in scales

snake

crocodile

See also: **cold-blooded, scale, vertebrate**

A snake is a **reptile**. A crocodile is also a **reptile**.

---

resistance (electricity)
/ri ZIS tuhns (i lek TRIS uh tee)/ (n.)

a material's ability to slow the flow of electricity

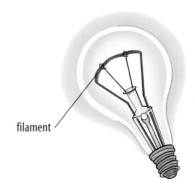

filament

See also: **circuit, conduct, electric current, electricity**

The filament is the part of a light bulb that glows. It glows because it has high **resistance**.

respiratory system
/RES puh ruh tor ee SIS tuhm/ (n.)

the group of organs involved
in breathing

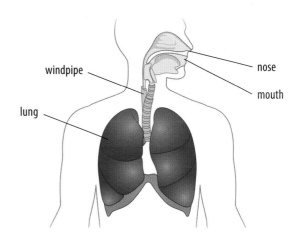

See also: **breathe, carbon dioxide,
lungs, oxygen**

The nose, mouth, windpipe, and lungs are part of the
respiratory system. These structures help you breathe.

---

response   /ri SPONS/ (n.)

a reaction to a stimulus

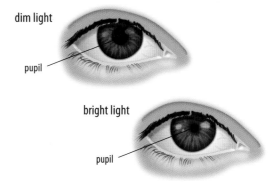

See also: **react, stimulus**

The pupil of the eye gets smaller as a **response**
to bright light.

---

retina   /RE tuh nuh/ (n.)

layers of cells at the back
of the eye that are sensitive
to light

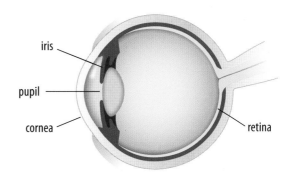

See also: **eye, lens, light, pupil**

The lens focuses light on the **retina**. The **retina** sends
signals to the brain.

157

reversible change
/ri VUR suh buhl chaynj/ (n.)

a change that can be undone

See also: **freezing point, irreversible change, melting point**

An ice cube melting to form water is a **reversible change**. The water can be frozen to form ice again.

revolve /ri VOLV/ (v.)

to move in a circle around something

An image showing Earth's orbit around the sun with labels September, December, sun, June, Earth, March.

See also: **orbit, planet, rotate**

Earth takes one year to **revolve** around the sun.

RNA /ar en AY/ (n.)

a molecule that has an important role in making proteins

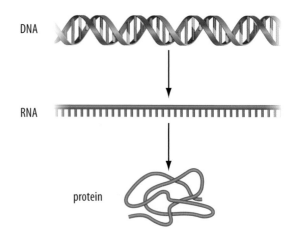

See also: **DNA, nucleic acid, protein, virus**

Information from a cell's DNA is passed to **RNA**. The **RNA** then helps make proteins.

rock cycle /rok SĪ kuhl/ (n.)

a natural process that creates
and changes rocks

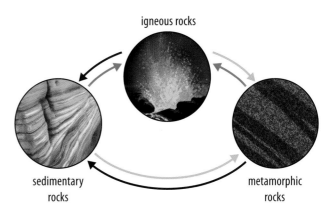

igneous rocks

sedimentary
rocks

metamorphic
rocks

See also: **cycle**

Each type of rock can change into another type
of rock in the **rock cycle**.

rocket /ROK it / (n.)

a device that is pushed up
into the air by hot gases

rocket

A **rocket** burns fuel
to create hot gases.
The hot gases escape
quickly and push the
rocket up into the air.

See also: **fuel**

rock strata /rok STRAY tuh/ (n.)

layers of rock

See also: **sedimentary rock**

Sedimentary rock may form **rock strata**.
The oldest layers are usually on the bottom.

159

root /root/ (n.)

the underground part of a plant

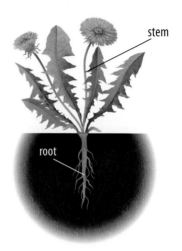

See also: **absorb, plant, soil, stem**

The **root** holds the plant in the soil. The **root** absorbs water and minerals from the soil.

rotate /ROH tayt/ (v.)

to turn on an axis

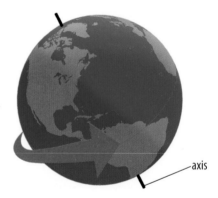

See also: **axis, revolve**

Earth takes one day to **rotate** on its imaginary axis.

ruler /ROO lur/ (n.)

a tool for measuring length

See also: **measure**

Inches and centimeters are often marked on a **ruler**.

salt /sawlt/ (n.)

a clear crystal used to flavor food

See also: **chemical, compound, crystal**

Salt is the chemical compound sodium chloride.

---

salt water /sawlt WAW tur/ (n.)

water that has salt dissolved in it

See also: **fresh water, marine, salt, water**

Earth's seas and oceans are **salt water**.

---

satellite /SA tuh līt/ (n.)

an object that orbits a planet or another object in space

See also: **moon, orbit, planet**

Communications **satellites** orbit Earth.

161

saturated /SA chuh ray ted/ (adj.)

not able to hold more

undissolved substance

See also: **dissolve, solute, solution, solvent**

A **saturated** solution will not dissolve any more of the substance.

---

scale /skayl/ (n.)

1. a tool used for measuring weight

See also: **balance, measure, weight**

A **scale** can measure weight in pounds or grams.

2. a ratio of the distance or size on a model to the real distance or size

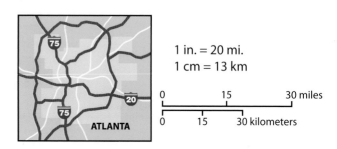

1 in. = 20 mi.
1 cm = 13 km

See also: **model**

A map has a **scale**. The **scale** relates the distance on the map to the real distance on Earth.

3. a type of plate that covers an animal's body

See also: **fish, reptile**

One type of **scale** covers reptiles.

scale model /skayl MOD uhl/ (n.)

an object that represents
a real object

See also: **model, scale**

A globe is a **scale model** of Earth.

scattered /SKA turd/ (adj.)

not moving in the same direction

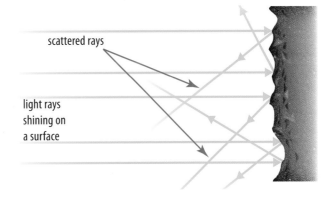

scattered rays

light rays
shining on
a surface

See also: **light, reflect, refract**

Light rays normally move in straight lines.
Light rays may become **scattered** if they hit
or pass through a material.

screw /skroo/ (n.)

a simple machine that can hold
two objects together

See also: **force, simple machine**

A **screw** is a tube with grooves. The grooves wind
around the tube in a spiral.

163

season /SEE zuhn/ (n.)

a time of the year with a certain type of weather

spring   summer
fall   winter

Each **season** has a different pattern of weather. Many parts of the world have four **seasons**: spring, summer, fall, and winter.

See also: **equinox, solstice, weather, year**

sediment /SE duh muhnt/ (n.)

particles of a solid that settle on the bottom of a liquid

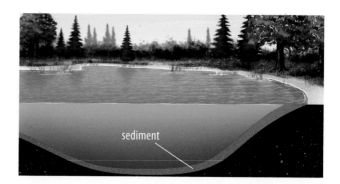

sediment

See also: **liquid, particle, sedimentary rock**

Sediment often settles on the bottom of a lake.

sedimentary rock
/se duh MEN tur ee rok/ (n.)

rock formed from layers of sediments

See also: **igneous rock, metamorphic rock, rock cycle, sediment**

Sandstone is a type of **sedimentary rock**. Layers of sand form sandstone.

seed /seed/ (n.)
a fertilized plant ovule

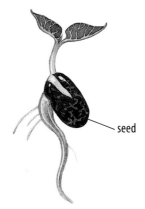

seed

See also: **fertilization, germinate, ovule, plant**

A new plant can grow from a **seed**. A **seed** contains food for the growing plant.

seedling /SEED ling/ (n.)
a young plant grown from a seed

See also: **plant, seed**

A **seedling** may develop after a seed is planted. A **seedling** may grow into an adult plant that can produce more seeds.

senses /SENS uhz/ (n.)
sight, hearing, smell, touch, and taste

touch   taste

hearing   smell   sight

See also: **ear, eye, nose, skin, tongue**

Our **senses** give us information about the world around us.

sepal /SEE puhl/ (n.)

an outer part of a flower

sepal

See also: **flower, petal**

A **sepal** helps protect the inner parts of the flower. **Sepals** are at the base of the flower petals.

series circuit
/SEER eez SUR kuht/ (n.)

an electric circuit that has only one conducting path

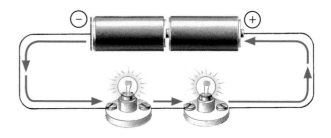

See also: **circuit, conduct, electricity, parallel circuit**

A battery-powered flashlight uses a **series circuit**.

sewage /SOO ij/ (n.)

waste carried away in a system of pipes

See also: **water**

Sewage includes waste water from bathrooms and kitchens.

sex /seks/ (n.)

male or female

male   female

See also: **gender, species**

The male and female **sex** of a species may look different.

shell /shel/ (n.)

a hard covering that protects
some animals

See also: **mollusk**

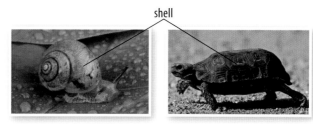

shell

A snail has a **shell**. A turtle also has a **shell**.

sieve /siv/ (n.)

a tool with small holes in it that
is used to separate materials

See also: **liquid, particle, solid**

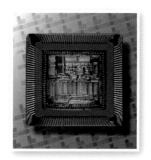

A **sieve** can separate larger particles from smaller
ones. It can also separate solids from liquids.

silicon (Si) /SIL uh kuhn/ (n.)

a common element
in Earth's crust

See also: **computer, crust, element**

Computer chips
contain **silicon**.

simple machine
/SIM puhl muh SHEEN/ (n.)

a basic device that makes
work easier

See also: **force, machine, work**

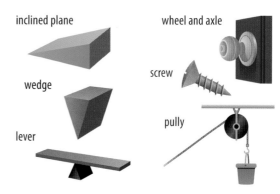

inclined plane   wheel and axle

wedge   screw

lever   pully

A lever is a **simple machine**. A wheel and axle is
also a **simple machine**.

167

sink /sink/ (v.)

to fall to the bottom of a liquid

Damaged ships sometimes **sink** to the bottoms of lakes and oceans.

See also: **buoyancy, density, float, liquid**

skeleton /SKE luh tuhn/ (n.)

a structure that supports or protects an organism's body

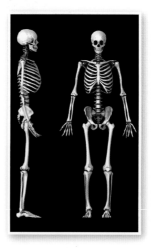

Humans have a **skeleton** that supports the body. The **skeleton** gives the body shape and helps it move.

See also: **bone, invertebrate, muscle, vertebrate**

skin /skin/ (n.)

layers of cells that cover an animal's body

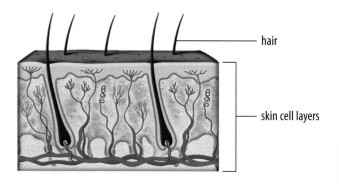

hair

skin cell layers

The human body is covered by **skin**. **Skin** has nerves that sense pain, pressure, and changes in temperature.

See also: **cell, nerve, senses**

skull /skuhl/ (n.)
the bones of the head

See also: **bone, brain, senses, skeleton**

The **skull** protects the brain and sense organs.

smell /smel/ (v.)
to sense an odor

See also: **nose, senses**

You use your nose to **smell** different odors.
A rose has a pleasant odor.

smog /smog/ (n.)
a type of air pollution

See also: **atmosphere, fossil fuel, pollution, react**

Smog is created when certain chemicals react with sunlight.

snow /snoh/ (n.)
frozen precipitation

See also: **crystal, freezing point, precipitation, water**

Snow is water frozen into ice crystals.

soil /soil/ (n.)

a mixture of particles of rock and organic matter

See also: **humus, organic matter, particle**

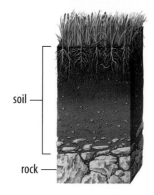

soil

rock

Soil makes up the top layer of Earth's surface. Plants grow in **soil**.

solar power /SOH lur POW ur/ (n.)

the energy of sunlight

solar panels

See also: **electricity, energy, heat, sun**

Solar power can heat houses. Solar panels change solar energy into electrical energy.

solar system
/SOH lur SIS tuhm/ (n.)

a sun and the planets that orbit it

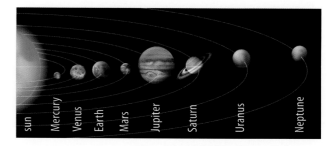

sun Mercury Venus Earth Mars Jupiter Saturn Uranus Neptune

See also: **orbit, planet, sun**

There are eight planets in our **solar system**.

solid /SOL id/ (n.)

a state of matter that has a definite shape and volume

Atoms are close together in a **solid**. Ice is the solid form of water.

See also: **atom, gas, liquid, state of matter**

solstice /SOL stis/ (n.)

the first day of summer and the first day of winter

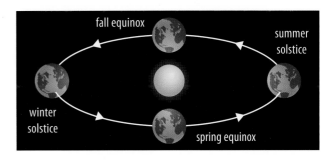

See also: **axis, equinox, hemisphere, season**

The summer **solstice** is the longest day of the year in the Northern Hemisphere. The winter **solstice** is the shortest day of the year in the Northern Hemisphere.

soluble /SOL yuh buhl/ (adj.)

able to be dissolved in another substance

Salt is **soluble** in water.

See also: **dissolve, insoluble, solution**

solute /SOL yoot/ (n.)

a substance that dissolves in another substance

See also: **dissolve, soluble, solution, solvent**

The color dye is the **solute** in each solution.

solution /suh LOO shuhn/ (n.)
a mixture of two or
more substances

See also: dissolve, soluble, solute, solvent

A solute and a solvent make up a **solution**.

solvent /SOL vuhnt/ (n.)
a substance in which another
substance is dissolved

solvent

Water is the **solvent** in
a saltwater solution.

See also: dissolve, soluble, solute, solution

sound waves /sownd wayvz/ (n.)
vibrations that travel in waves
through matter

A vibrating object creates **sound waves**. **Sound
waves** can move through air, water, or other materials.

See also: ear, hear, vibration, wave

space /spays/ (n.)
the part of the universe outside
Earth's atmosphere

space

Earth

See also: atmosphere, galaxy, satellite,
solar system

Space includes the solar system
and distant galaxies.

species /SPEE sheez/ (n.)

a group of organisms that can produce offspring

See also: **genus, offspring, reproduction (sexual)**

Panthera tigris

Individuals of a **species** can mate and produce offspring. A tiger belongs to the **species** *Panthera tigris*.

---

spectrum /SPEK truhm/ (n.)

range

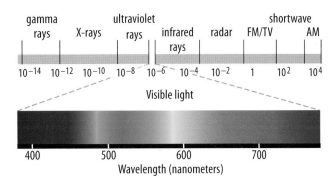

The visible light **spectrum** includes the colors of the rainbow.

See also: **light, prism, wavelength**

---

speed /speed/ (n.)

the distance an object moves over a period of time

See also: **rate, velocity**

Speed is the rate at which something moves. It is measured in units like kilometers per hour.

---

sperm /spurm/ (n.)

the male gamete

sperm

egg

See also: **egg, embryo, fertilization, gamete, ovum**

Male animals produce **sperm**. **Sperm** fertilize an egg to make an embryo.

spine /spīn/ (n.)

the row of connected bones
down the middle of the back

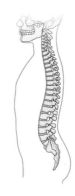

The **spine** helps support the body.

See also: **backbone, vertebra, vertebrate**

spore /spor/ (n.)

a cell that some organisms use
to reproduce

A single **spore** can develop
into a new organism. This
fungus is releasing **spores**.

See also: **meiosis, mitosis, reproduction
(asexual), reproduction (sexual)**

spring /spring/ (n.)

a metal coil

A **spring** returns to its original
shape after it is pressed down
or stretched.

See also: **energy**

spring balance
/spring BAL uhns/ (n.)

a tool that uses a spring
to measure weight

A **spring balance** has a spring
inside. A weight on the hook
pulls the spring down.

See also: **balance, scale, spring, weight**

stamen /STAY muhn/ (n.)

the male reproductive part of a flower

filament —stamen
anther

See also: **anther, filament, flower, pollen**

The anther and filament make up the **stamen**.

---

star /star/ (n.)

a large ball of gas that produces radiation

A **star** may appear very bright through a telescope.

See also: **gas, radiation, sun, telescope**

---

state of matter
/stayt uhv MAT ur/ (n.)

solid, liquid, or gas

solid   liquid   gas

See also: **gas, liquid, matter, solid**

Each **state of matter** has different properties. The molecules are arranged in different ways.

---

stem /stem/ (n.)

the part of a plant that holds the leaves and flowers

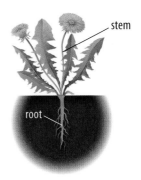

stem

root

See also: **phloem, plant, tissue, xylem**

The **stem** has special tissues. These tissues carry water and food for the plant.

175

sterile /STER uhl/ (adj.)

without microorganisms growing on it

A doctor uses **sterile** gloves and other equipment. This helps prevent infections.

See also: **infect, microorganisms**

stethoscope /STE thuh skohp/ (n.)

a tool for listening to the heart and lungs

stethoscope

A doctor uses a **stethoscope**.

See also: **heart, lung**

stigma /STIG muh/ (n.)

the top of the female reproductive part of a flower

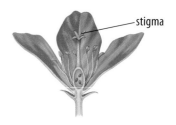

stigma

The **stigma** is sticky so pollen gets stuck on it. Pollen then travels down the tube to the ovary.

See also: **flower, ovary, pollen**

stimulus /STIM yuh luhs/ (n.)

something that causes a reaction

You move your hand away if you touch something very hot. Heat is the **stimulus** that causes you to move.

See also: **reaction, reflex**

stoma (plural stomata)
/STOH muh/ (n.)

a small hole in a leaf that lets gases in and out

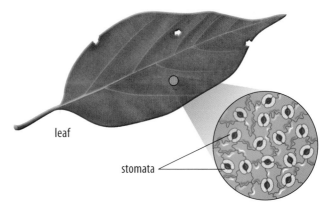

leaf

stomata

See also: **leaf, photosynthesis**

Each **stoma** can open and close. A leaf has many **stomata**.

stomach /STUHM uhk/ (n.)

the organ that receives chewed food through the esophagus

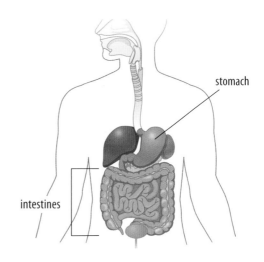

stomach

intestines

See also: **digestive system, esophagus, organ**

The **stomach** is part of the digestive system. Strong acid in the **stomach** breaks down food.

stopwatch /STOP wahch/ (n.)

a watch that can be started and stopped quickly

See also: **measure**

Use a **stopwatch** to measure the amount of time it takes to do something.

177

stream /streem/ (n.)
a small body of flowing water

See also: **habitat, surface water, water**

A **stream** is a habitat for many organisms.

stress /stres/ (n.)
a type of harmful stimulus

See also: **stimulus**

Too much work can cause **stress**. Injury or illness can also put **stress** on a body.

sugar /SHUG ur/ (n.)
a type of carbohydrate that tastes sweet

See also: **carbohydrate**

Plants make most **sugar**. **Sugar** can sweeten food.

sun /suhn/ (n.)
the star at the center of our solar system

See also: **radiation, solar power, solar system, star**

The **sun** produces radiation. The radiation heats and lights Earth and other planets.

sundial /SUHN dī uhl/ (n.)

a tool that uses the position
of a shadow to tell the time

See also: **sun, technology**

The shadow on the **sundial** shows the time.
The shadow's position changes as the sun's position
in the sky changes.

surface water
/SUR fis WAW tur/ (n.)

water found above ground

See also: **groundwater, ocean,
stream, water**

Rivers, lakes, seas, and oceans contain **surface water**.

survive /sur VĪV/ (v.)

to stay alive

See also: **environment, organism**

Some organisms **survive** in very difficult
environments like the desert.

sweat /swet/ (n.)

salty water produced by glands in the skin

See also: **evaporation, gland, skin**

Sweat evaporates and cools the body.

switch /swich/ (n.)

a device that stops and starts the flow of electricity

See also: **circuit, electric current, electricity**

A **switch** is part of an electrical circuit.

symbiosis /sim bee OH suhs/ (n.)

a close relationship between organisms of different species

See also: **organism, parasite, species**

Symbiosis may benefit one or more of the organisms.

synthetic /sin THE tik/ (adj.)

made by people, not nature

See also: **natural**

Plastic is a **synthetic** material.

taiga /TĪ guh/ (n.)

a biome with a cool climate and coniferous forests

See also: **biome, climate, conifer, evergreen**

The **taiga** covers northern regions around the globe.

technology /tek NOL uh jee/ (n.)

the helpful use of science and other knowledge

Pencils are an example of **technology**. A cell phone is **technology** too.

telescope /TEL uh skohp/ (n.)

a device that is used to make distant objects appear closer

See also: **lens, light, magnify, planet**

A **telescope** may use lenses to focus light. Some **telescopes** use mirrors or both mirrors and lenses.

181

temperature
/TEM pur uh chur/ (n.)

the measure in degrees
of how hot something is

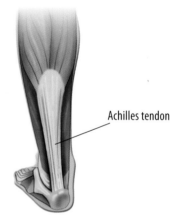

Monday	
High	**30°C**
Low	**19°C**

The **temperature** may reach 30°C on Monday.

See also: **Celsius, Fahrenheit, thermometer**

---

tendon /TEN duhn/ (n.)

tissue that connects muscle
to bone

Achilles tendon

The Achilles **tendon** is the **tendon** at the back
of your ankle.

See also: **bone, muscle, tissue**

---

theory /THEE ur ee/ (n.)

a scientific explanation based
on a large amount of evidence

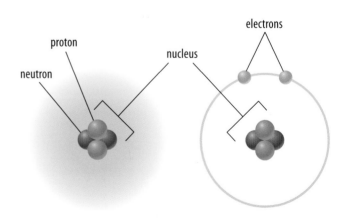

electrons

proton

nucleus

neutron

See also: **atom, evidence, evolution, hypothesis**

Atoms make up all matter according to atomic **theory**.

thermometer
/thur MOM uh tur/ (n.)

a tool used to measure temperature

See also: **Celsius, Fahrenheit, temperature**

You can use a **thermometer** to take a person's temperature.

thorax /THOR aks/ (n.)

the region of an animal's body between the head and abdomen

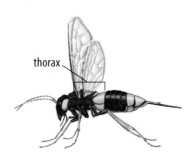

thorax

An insect's legs and wings are attached to its **thorax**.

See also: **animal, head, insect**

throat /throht/ (n.)

the body part that connects the mouth and nose to internal organs

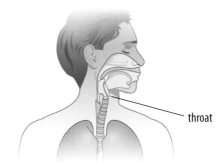

throat

See also: **esophagus, mouth, nose**

The **throat** contains the esophagus.

thunder /THUN dur/ (n.)

the sound produced by lightning

First you see lightning. Then you hear **thunder**.

See also: **electricity, light, lightning, sound waves**

183

tide /tīd/ (n.)
a change in sea level caused by the pull of the moon's gravity

low tide   high tide

See also: **gravity, moon, ocean**

Ocean water is highest at high **tide**. The water is lowest at low **tide**.

tissue /TI shoo/ (n.)
a group of cells that work together to perform a particular function

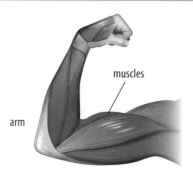

muscles

arm

Muscle **tissue** makes up your muscles.

See also: **cell, organ**

tongue /tuhng/ (n.)
a mouth part that helps with eating, tasting, and speaking

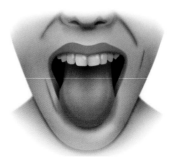

Your **tongue** helps you pronounce words.

See also: **digestive system, food, mouth, tooth**

tooth (plural **teeth**) /tooth/ (n.)
one of the hard structures found in the mouth and used for chewing

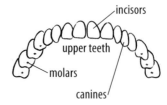

incisors

upper teeth

molars

canines

Each kind of **tooth** has a special role in biting or chewing food.

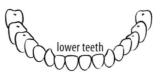

lower teeth

See also: **digestive system, mouth, tongue**

tornado /tor NAY doh/ (n.)

a storm with strong, destructive winds

A funnel cloud forms during a **tornado**.

See also: **cloud, hurricane, weather**

toxic /TOK sik/ (adj.)

able to cause illness or death

See also: **chemical, flammable, radioactive**

⚠ CAUTION

HIGHLY TOXIC CHEMICAL

Some chemicals are **toxic**. It is important to read warning labels carefully.

trait /trayt/ (n.)

a characteristic that is passed from parent to offspring

See also: **DNA, gene, heredity, offspring**

Eye color is an example of a **trait** that you can see.

translucent
/tranz LOO suhnt/ (adj.)

allowing some, but not all, light to pass through

You can see only a blurred image through a **translucent** material.

See also: **light, opaque, transparent**

transparent /tranz PAIR uhnt/ (adj.)

allowing almost all light
to pass through

See also: **gravity, moon, ocean**

A clear glass window is **transparent**.

transpiration

/tran spuh RAY shuhn/ (n.)

the evaporation of water from
the surface of a plant

See also: **evaporation, plant**

Plants can lose large amounts of water
through **transpiration**.

tree /tree/ (n.)

a large woody plant with a trunk

trunk

See also: **deciduous, evergreen**

A deciduous **tree** loses its leaves in fall.
An evergreen **tree** keeps its leaves.

tropical /TROP uh kuhl/ (adj.)

characteristic of regions
near Earth's equator

See also: **biodiversity, biome, equator, rain forest**

Tropical areas have many different kinds of plants and animals.

trough /trawf/ (n.)

the lowest point of a wave

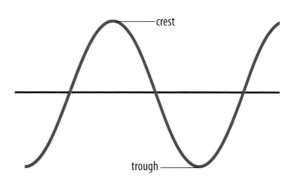

See also: **amplitude, crest, wave, wavelength**

The **trough** of the wave is opposite the crest of the wave.

tundra /TUHN druh/ (n.)

a biome with a cold climate
and no trees

See also: **biome, climate, taiga**

Trees are not able to grow in the cold climate of the **tundra**.

ultrasound /UHL truh sownd/ (n.)

a sound with a frequency too
high for humans to hear

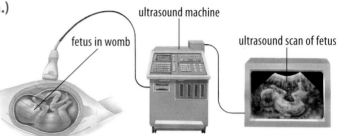

ultrasound machine

fetus in womb

ultrasound scan of fetus

See also: **echo, fetus, frequency,
sound waves**

Some medical tools use **ultrasound**. A machine
turns **ultrasound** echoes into an image.

ultraviolet (UV) light

/uhl truh VĪ uh lit līt/ (n.)

a type of light that carries more
energy than visible light

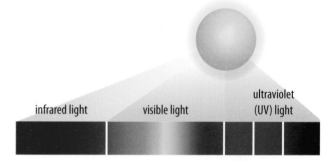

infrared light

visible light

ultraviolet
(UV) light

See also: **infrared light, light, radiation, sun**

You cannot see **ultraviolet light**. **UV light**
from the sun can damage skin.

umbilical cord

/uhm BIL uh kuhl kord/ (n.)

the structure that connects a
mammal fetus to the placenta

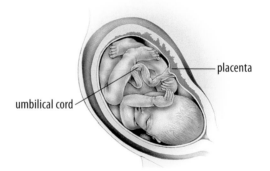

placenta

umbilical cord

See also: **embryo, fetus, mammal, placenta**

The **umbilical cord** carries blood to a developing
mammal. It carries away wastes.

unicelluar
/yoo nuh SEL yuh lur/ (adj.)
made of a single cell

bacteria   protists

See also: **bacterium, multicellular, protist**

Bacteria and many protists are **unicellular** organisms.

universe /YOO nuh vurs/ (n.)
everything that exists

See also: **Big Bang Theory, galaxy, solar system**

The **universe** contains many galaxies. Earth is in the Milky Way galaxy.

uterus /YOO tur uhs/ (n.)
the structure in female mammals that offspring develop in

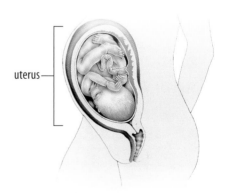

uterus—

See also: **gestation, mammal, placenta**

Layers of muscles make up the **uterus**. The **uterus** grows larger as the developing offspring grows.

Vv

vaccine /vak SEEN/ (n.)
a medicine that helps the body become resistant to a disease

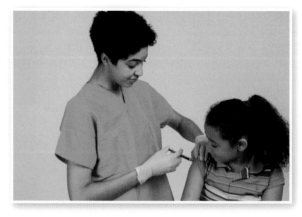

See also: **disease, immune system, immunize, virus**

A **vaccine** causes the body's immune system to respond. The immune system prepares to fight a certain disease.

vacuum /VAK yoom/ (n.)
a space that contains no matter

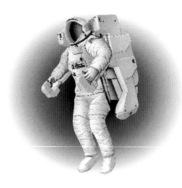

See also: **matter, space**

Outer space is mostly a **vacuum**. Humans need to wear air tanks in outer space.

valley /VA lee/ (n.)
a U-shaped or V-shaped landform

See also: **glacier, mountain, stream**

A river or stream forms a V-shaped **valley**.

190

valve /valv/ (n.)

a structure that controls
the movement of fluid

See also: **aorta, blood, heart**

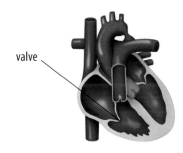

valve

A heart **valve** allows blood to flow in only one
direction. There are four **valves** in the heart.

---

vegetation
/ve juh TAY shuhn/ (n.)

all of the plants that grow
in an area

See also: **biome, plant**

The **vegetation** is different in each biome.

---

vein /vayn/ (n.)

a blood vessel that carries blood
back to the heart

See also: **aorta, artery, capillary, heart**

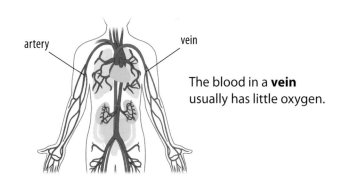

artery   vein

The blood in a **vein**
usually has little oxygen.

---

velocity /vuh LOS uh tee/ (n.)

the speed and direction
of an object's movement

See also: **accelerate, inertia, speed**

The jet's **velocity** is 500 mph north.

vertebra (plural vertebrae)
/VUR tuh bruh/ (n.)

one of the bones that make up
the backbone

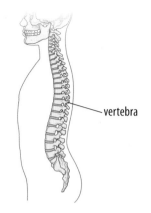
vertebra

See also: **backbone, spine, vertebrate**

You can feel bumps if you run your hand down the
middle of your back. Each bump is a **vertebra**.

vertebrate   /VUR tuh brit/ (n.)
an animal with a backbone

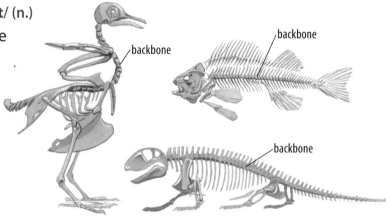
backbone

backbone

backbone

See also: **backbone, spine, vertebra**

A bird is an example of a **vertebrate**. Fish,
amphibians, reptiles, and mammals are
vertebrates too.

vibration   /vī BRAY shuhn/ (n.)
an up-and-down or
back-and-forth movement

See also: **sound waves, wave**

Vibration of air produces sound in wind instruments.

virus /VĪ ruhs/ (n.)

a microscopic structure with DNA or RNA; it attacks and uses cells to make copies of itself

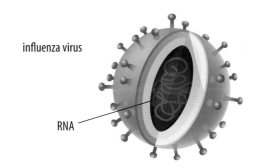

influenza virus

RNA

See also: **DNA, HIV, infect, influenza, RNA**

A **virus** can cause many different illnesses. **Viruses** cause the common cold and influenza.

vitamin /VĪ tuh muhn/ (n.)

a substance needed in very small amounts to stay healthy

See also: **diet, molecule**

Oranges contain a **vitamin** called **vitamin** C.

volcano /vol KAY noh/ (n.)

a mountain that is built up from layers of cooled lava

See also: **igneous rock, lava, magma, mountain**

The lava from a **volcano** cools into igneous rock.

volume /VOL yuhm/ (n.)

the amount of space an object takes up

See also: **graduated cylinder, mass**

You can use a graduated cylinder to measure the **volume** of a liquid.

waning /WAY ning/ (v.)
decreasing in size

See also: **Earth, moon, orbit, waxing**

The **waning** moon occurs between the full moon and the new moon. We see less sunlight reflected from it.

---

warm-blooded
/WAWRM bluhd id/ (adj.)
having a body temperature that always stays the same

Birds are **warm-blooded**. Mammals are **warm-blooded** too.

See also: **birds, cold-blooded, mammals**

---

water /WAW tur/ (n.)
a liquid made of hydrogen and oxygen

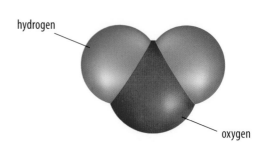

hydrogen

oxygen

See also: **liquid, molecule, solid, water vapor**

Water can freeze into ice or evaporate into **water** vapor. Living things need **water** to survive.

water cycle /WAW tur SĪ kuhl/ (n.)

the movement of water between Earth and Earth's atmosphere

See also: **condensation, evaporation, precipitation**

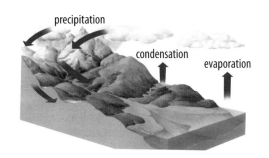

Water is constantly moving through the **water cycle**.

watershed /WAW tur shed/ (n.)

a region of land from which surface water drains

See also: **groundwater, surface water**

Water runs downhill from all parts of a **watershed**.

water vapor
/WAW tur VAY pur/ (n.)

water in gas form

See also: **gas, liquid, solid, state of matter**

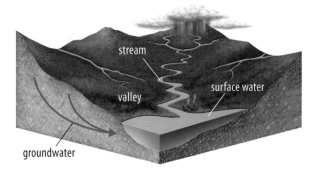

When water evaporates it becomes **water vapor**. **Water vapor** is invisible.

watt /waht/ (n.)

a unit for the measurement of electrical energy

Some light bulbs use more energy than others. This light bulb is made to use 15 **watts**.

See also: **ampere, electric current, electricity**

wave /wayv/ (n.)

an up-and-down or back-and-forth movement that transfers energy

crest

trough

See also: **amplitude, light, sound waves, wavelength**

A water **wave** transfers energy through the ocean. Sound and light travel in **waves** too.

wavelength /wayv lengkth/ (n.)

the distance between two crests or troughs of a wave

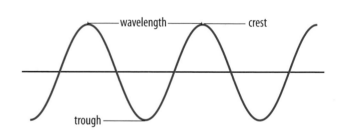

wavelength ———— crest

trough

See also: **crest, frequency, pitch, trough**

Each color is a particular **wavelength** of light.

waxing /WAKS ing/ (v.)

increasing in size

See also: **Earth, moon, orbit, waning**

The **waxing** moon occurs between the new moon and the full moon. We see more sunlight reflected from it.

weather   /WETH ur/ (n.)

the temperature, moisture, and wind conditions at a particular place and time

See also: **atmosphere, climate, precipitation, temperature**

The **weather** today will be warm and cloudy with a chance of rain.

---

weathering   /WETH ur ing/ (n.)

the set of processes that breaks down rock over time

See also: **erosion, soil**

Ice, water, and wind can all cause **weathering**.

---

wedge   /wej/ (n.)

a simple machine that can be used to split things apart

See also: **force, simple machine**

wedge

The end of an axe is a **wedge**.

---

weight   /wayt/ (n.)

the pull of gravity on an object

See also: **gravity, mass, newton, spring balance**

3 kg

A spring balance can measure **weight**.

wheel and axle
/weel and AK suhl/ (n.)

a simple machine that can
be used to lift or move things

See also: **force, simple machine**

A doorknob is an example of a **wheel and axle**.

---

wing /wing/ (n.)

a structure usually used for flying

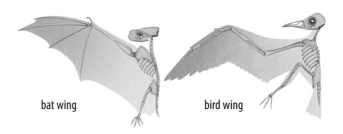

bat wing   bird wing

See also: **aerodynamic, bird, insect**

A bat's **wing** is made of skin. A bird's **wing** has
feathers. Insects and airplanes have **wings** too.

---

wire /wīr/ (n.)

a strand of metal

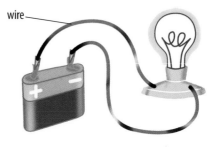

wire

See also: **circuit, electric current, metal**

An electric current runs through the **wire** in a circuit.

---

work /wurk/ (n.)

a use of force that makes
something move or change

See also: **energy, force, simple machine**

The crane is doing **work** when it lifts a load.

Xx

X-ray /EKS ray/ (n.)

a form of energy; it is often used in medical technology

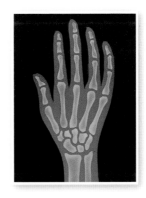

See also: **bone, energy, technology, tissue**

An **X-ray** can pass easily through soft tissue. **X-rays** can be used to make images of bones.

xylem /ZĪ luhm/ (n.)

plant tissue that transports water and minerals

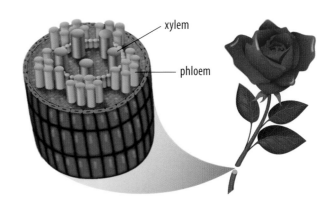

xylem

phloem

See also: **phloem, plant, tissue, transpiration, tree**

There is **xylem** in the stem of a flower.

Yy

year /yeer/ (n.)

the amount of time it takes a planet to orbit the sun

See also: **Earth, orbit, sun**

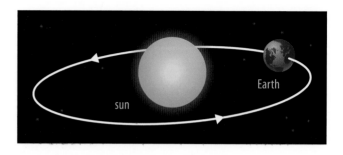

A **year** on Earth is just over 365 days.

yeast /yeest/ (n.)

a single-celled fungus

before

after

See also: **eukaryote, fermentation, fungus, unicellular**

People use **yeast** to make bread rise.

yolk /yohk/ (n.)

the yellow part of an egg

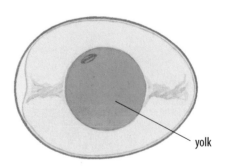

yolk

See also: **egg, embryo**

The **yolk** is the food supply for a developing bird embryo.

zoology /zoh OL uh jee/ (n.)
the study of animals

See also: **biology, botany, evolution**

Zoology includes many different topics. It includes the study of animal structure, evolution, and behavior.

zygote /ZĪ goht/ (n.)
the cell that results from the joining of a sperm and egg

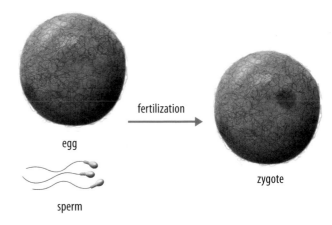

egg

fertilization

zygote

sperm

See also: **embryo, fertilization, gamete, reproduction (sexual)**

A **zygote** is the cell that forms after fertilization. The **zygote** develops into an embryo.

Life Science

abdomen	bone	decompose
adaptation	botany	decomposer
adult	brain	diabetes
age	breathe	diaphragm
algae	breed	diet
alimentary canal	bud	digestion
alive	bulb	digestive system
allergy	cactus	dinosaur
amphibian	caffeine	disease
anatomy	camouflage	DNA
ancestor	cancer	dominant trait
anesthetic	capillary	ear
animal	carnivore	ecology
antennae	caterpillar	ecosystem
anther	cell	egg
antibiotic	cellular respiration	embryo
antiseptic	cereal	endangered species
anus	cerebellum	enzyme
aorta	cerebrum	esophagus
arachnid	chlorophyll	eukaryote
artery	chloroplast	evergreen
arthropod	chromosome	evolution
asthma	chrysalis	evolve
backbone	circulatory system	exercise
bacterium	claw	exhale
bark	cocoon	extinct
beak	cold-blooded	eye
beetle	compost	feather
binocular vision	conifer	female
biodegradable	consumer	fertilization
biodiversity	coral	fetus
biology	cornea	filament
biome	cranium	fin
biped	crustacean	fish
bird	dead	flower
bladder	deaf	food chain
blind	decay	food web
blood	deciduous	fossil

fossil record

fruit

fungus

gamete

gender

gene

genitals

genotype

genus

germ

germinate

gestation

gill

gland

grass

grassland

grow

habit

habitat

hair

head

healthy

hear

heart

herbivore

hibernate

HIV

honey

hostile

human

humerus

immune system

immunize

incubate

index fossil

indigenous

infect

influenza

inhale

insect

instinct

intestine

invertebrate

iris

jaw

joint

kidney

kingdom

larva

leaf

lens

life cycle

life processes

limb

limiting factor

liver

lung

male

mammal

mature

meiosis

membrane

menstruation

metamorphosis

microorganism

milk

mitochondrion

mitosis

mold

mollusk

moss

mouth

mucus

multicellular

muscle

mushroom

mutant

mutation

natural

natural selection

neck

nerve

nervous system

nicotine

nocturnal

non-living

nose

nucleus

nutrient

offspring

omnivore

organ

organic matter

organism

ovary

ovule

ovum

pancreas

parasite

petal

phenotype

phloem

photosynthesis

pistil

placenta

plankton

plant

pneumonia

poison

pollen

pollinate

population

predator

prey

producer
prokaryote
protist
puberty
pulse
pupa
pupil
react
recessive trait
red blood cell
reflex
reproduction (asexual)
reproduction (sexual)
reproductive system
reptile
respiratory system
response
retina
root
rot
scale
seed
seedling
senses
sepal

sex
shell
skeleton
skin
skull
smell
species
sperm
spine
spore
stamen
stem
sterile
stigma
stimulus
stoma
stomach
stress
survive
sweat
symbiosis
tendon
testis
thorax
throat

tissue
tongue
tooth
trait
transpiration
tree
umbilical cord
unicellular
uterus
vaccine
valve
vegetation
vein
vertebra
vertebrate
virus
vitamin
warm-blooded
wing
xylem
yeast
yolk
zoology
zygote

Earth and Space Science

acid rain
air
air mass
air pressure
ancient
annual
aquifer
asteroid

astronomy
atmosphere
axis
barometer
Big Bang theory
black hole
canyon
carbon cycle

climate
climate change
cloud
coal
comet
conservation
constellation
core

corona
crater
crust
crystal
day
delta
desert
dew
diamond
Earth
earthquake
eclipse
environment
equator
equinox
erosion
fault
fog
fossil fuel
fresh water
frost
fusion
galaxy
geology
glacier
gravel
greenhouse effect
groundwater
hardness
hemisphere
horizon
humidity
humus
hurricane
ice
igneous rock
lava

light year
lightning
limestone
lunar
magma
mantle
marine
metamorphic rock
meteorite
mineral
moon
mountain
night
nitrogen cycle
non-renewable energy
nuclear power
ocean
oil
orbit
ore
ozone
peat
planet
plate tectonics
plateau
pollution
precipitation
rain forest
rainfall
reef
renewable energy
revolve
rock cycle
rock strata
rotate
salt water
satellite

season
sediment
sedimentary rock
sewage
smog
snow
soil
solar power
solar system
solstice
space
star
stream
sun
surface water
taiga
thunder
tide
tornado
toxic
tropical
tundra
universe
vacuum
valley
volcano
waning
water
water cycle
water vapor
watershed
waxing
weather
weathering
year

Chemistry

acid
alloy
aluminum
atom
base
boiling point
burn
calcium
calorie
carbohydrate
carbon
carbon dioxide
chemical
chemical change
chemistry
compound
concentration
copper
corrosion
density
diffusion
dilute
dissolve
distill
electron
element
explosion
fat
fermentation

fission
flame
flammable
freezing point
fuel
gas
glucose
hydrogen
ice
impermeable
insoluble
ion
iron
irreversible change
lipid
liquid
mass
matter
melting point
metal
mixture
mole
molecule
neutron
nitrogen
nucleic acid
nucleus
osmosis
oxidation

oxide
oxygen
particle
periodic table
permeable
pH
phase (change)
physical change
property
protein
proton
radioactive
react
reactant
reaction
reduction
reversible change
salt
saturated
silicon
solid
soluble
solute
solution
solvent
state of matter
sugar
water vapor

Physics

absorb
accelerate
ampere
amplitude
buoyancy
center of gravity
color
concave
condensation
conduct
conduction
convection
convex
crest
echo
electric charge
electric current
electricity
energy
evaporation
filter
float
fluid
force
frequency
friction

fulcrum
gravity
heat
image
inertia
infrared light
insulate
kilogram
kinetic energy
lens
lever
light
magnet
magnetic pole
magnetism
magnify
mirror
momentum
net force
newton
opaque
physics
pitch
potential energy
pressure
prism

radiant
radiation
rainbow
react
reflect
refract
resistance
scattered
sink
sound waves
spectrum
speed
temperature
translucent
transparent
trough
ultraviolet (UV) light
vacuum
valve
velocity
vibration
wave
wavelength
weight
work
X-ray

Science as Inquiry

balance	litmus paper	range
classify	look	rate
cycle	maximum	real image
data	measure	record
dissect	minimum	ruler
evidence	mix	scale
experiment	model	scale model
graduated cylinder	observe	sieve
hypothesis	plan	spring balance
investigate	predict	stopwatch
listen	rain gauge	theory

Engineering, Technology, and Society

aerodynamic	glass	rocket
air resistance	inclined plane	satellite
amplify	kilowatt	screw
automatic	laser	series circuit
battery	lens	simple machine
biodegradable	lever	solar power
Celsius	machine	spring
circuit	magnifying glass	stethoscope
compass (magnetic)	manual	submarine
computer	medicine	sundial
drug	microscope	switch
electromagnet	mirror	synthetic
engine	motor	technology
Fahrenheit	nylon	telescope
fiber optics	parachute	thermometer
filament	parallel circuit	ultrasound
fulcrum	pendulum	watt
fuse	plastic	wedge
gear	prism	wheel and axle
generator	pulley	wire

Converting Units

Unit	Conversion Factor
centimeter	× 0.394 = inches
meter	× 3.281 = feet
kilometer	× 0.621 = miles
gram	× 0.035 = ounces
kilogram	× 2.205 = pounds
degrees Celsius	(°C × 1.8) + 32 = degrees Fahrenheit
milliliter (cubic centimeter)	× 0.034 = fluid ounces
liter	× 0.264 = gallons
joule	× 0.239 = calories
inch	× 2.54 = centimeters
foot	× 0.305 = meters
mile	× 1.609 = kilometers
ounce	× 28.35 = grams
pound	× 0.454 = kilograms
degrees Fahrenheit	(°F − 32) × 0.556 = degrees Celsius
fluid ounce	× 29.574 = milliliters (cubic centimeters)
gallon	× 3.785 = liters
calorie	× 4.184 = joules

The Periodic Table of Elements

1 **H** HYDROGEN																	2 **He** HELIUM
3 **Li** LITHIUM	4 **Be** BERYLIUM											5 **B** BORON	6 **C** CARBON	7 **N** NITROGEN	8 **O** OXYGEN	9 **F** FLUORINE	10 **Ne** NEON
11 **Na** SODIUM	12 **Mg** MAGNESIUM											13 **Al** ALUMINIUM	14 **Si** SILICON	15 **P** PHOSPHORUS	16 **S** SULFUR	17 **Cl** CHLORINE	18 **Ar** ARGON
19 **K** POTASSIUM	20 **Ca** CALCIUM	21 **Sc** SCANDIUM	22 **Ti** TITANIUM	23 **V** VANADIUM	24 **Cr** CHROMIUM	25 **Mn** MANGANESE	26 **Fe** IRON	27 **Co** COBALT	28 **Ni** NICKEL	29 **Cu** COPPER	30 **Zn** ZINC	31 **Ga** GALLIUM	32 **Ge** GERMANIUM	33 **As** ARSENIC	34 **Se** SELENIUM	35 **Br** BROMINE	36 **Kr** KRYPTON
37 **Rb** RUBIDIUM	38 **Sr** STRONTIUM	39 **Y** YTTRIUM	40 **Zr** ZIRCONIUM	41 **Nb** NIOBIUM	42 **Mo** MOLYBDENUM	43 **Tc** TECHNETIUM	44 **Ru** RUTHENIUM	45 **Rh** RHODIUM	46 **Pd** PALLADIUM	47 **Ag** SILVER	48 **Cd** CADMIUM	49 **In** INDIUM	50 **Sn** TIN	51 **Sb** ANTIMONY	52 **Te** TELLURIUM	53 **I** IODINE	54 **Xe** XENON
55 **Cs** CAESIUM	56 **Ba** BARIUM	57-71 LANTHANOIDS	72 **Hf** HAFNIUM	73 **Ta** TANTALUM	74 **W** TUNGSTEN	75 **Re** RHENIUM	76 **Os** OSMIUM	77 **Ir** IRIDIUM	78 **Pt** PLATINUM	79 **Au** GOLD	80 **Hg** MERCURY	81 **Tl** THALLIUM	82 **Pb** LEAD	83 **Bi** BISMUTH	84 **Po** POLONIUM	85 **At** ASTATINE	86 **Rn** RADON
87 **Fr** FRANCIUM	88 **Ra** RADIUM	89-103 ACTINOIDS	104 **Rf** RUTHERFORDIUM	105 **Db** DUBNIUM	106 **Sg** SEABORGIUM	107 **Bh** BOHRIUM	108 **Hs** HASSIUM	109 **Mt** MEITNERIUM	110 **Ds** DARMSTADTIUM	111 **Rg** ROENTGENIUM	112 **Cn** COPERNICIUM	113 **Uut** UNUNTRIUM	114 **Uuq** UNUNQUADIUM	115 **Uup** UNUNPENTIUM	116 **Uuh** UNUNHEXIUM	117 **Uus** UNUNSEPTIUM	118 **Uuo** UNUNOCTIUM

57 **La** LANTHANUM	58 **Ce** CERIUM	59 **Pr** PRAESODYMIUM	60 **Nd** NEODYMIUM	61 **Pm** PROMETHIUM	62 **Sm** SAMARIUM	63 **Eu** EUROPIUM	64 **Gd** GADOLINIUM	65 **Tb** TERBIUM	66 **Dy** DYSPROSIUM	67 **Ho** HOLMIUM	68 **Er** ERBIUM	69 **Tm** THULIUM	70 **Yb** YTTERBIUM	71 **Lu** LUTETIUM
89 **Ac** ACTINIUM	90 **Th** THULIUM	91 **Pa** PROTACTINIUM	92 **U** URANIUM	93 **Np** NEPTUNIUM	94 **Pu** PLUTONIUM	95 **Am** AMERICIUM	96 **Cm** CURIUM	97 **Bk** BERKELIUM	98 **Cf** CALIFORNIUM	99 **Es** EINSTEINIUM	100 **Fm** FERMIUM	101 **Md** MENDELEVIUM	102 **No** NOBELIUM	103 **Lr** LAWRENCIUM

Some Scientific Equations

Quantity	Equation	Standard Units
acceleration	acceleration (a) = force (F) ÷ mass (m) $$a = \frac{F}{m}$$	$\frac{\text{meters}}{\text{second}^2}$ ($\frac{m}{s^2}$)
density	density (d) = mass (m)/volume (V) $$d = \frac{m}{V}$$	$\frac{\text{kilogram}}{\text{meter}^3}$ ($\frac{kg}{m^3}$) $\frac{\text{gram}}{\text{centimeter}^3}$ ($\frac{g}{cm^3}$)
force	force (F) = mass (m) × acceleration (a) $$F = m \times a$$	newton (N)
momentum	momentum (p) = mass (m) × velocity (v) $$p = m \times v$$	kilogram × $\frac{\text{meters}}{\text{second}}$ (kg × $\frac{m}{s}$)
pressure	pressure (p) = force (F) ÷ area (A) $$p = \frac{F}{A}$$	pascal (Pa)
velocity	velocity (v) = distance (d) ÷ time (t) $$v = \frac{d}{t}$$	$\frac{\text{meters}}{\text{second}}$ ($\frac{m}{s}$)
weight	weight (W) = mass (m) × acceleration due to gravity (g) $$W = m \times g$$	newton (N)
work	work (W) = force (F) × distance (d) $$W = F \times d$$	joule (J)

Some Common Scientific Quantities

Quantity	Amount	Symbol Key
standard gravity (g)	$g = \frac{9.806 \text{ m}}{s^2}$	standard gravity (g); meter (m); second (s)
density of water	$1 \frac{g}{m^3}$	gram (g); meter (m)
normal freezing point of water	0°C (32°F)	Celcius (C); Fahrenheit (F)
normal boiling point of water	100°C (212°F)	Celcius (C); Fahrenheit (F)
standard atmospheric pressure (atm)	1 atm = 101,325 Pa	standard atmospheric pressure (atm); pascal (Pa)
speed of light (c) in a vacuum	299,792,458 $\frac{m}{s}$	meter (m); second (s)
speed of sound at sea level	340.290 $\frac{m}{s}$	meter (m); second (s)

Greek and Latin Word Roots

Prefix	Meaning	Example
aero-	air	aerodynamic
agri-, agro-	field	agriculture
alt-	high	altitude
amphi-	both	amphibian
anti-	against	antibiotic
astro-	star	astronaut
auto-	self	automatic
baro-	pressure	barometer
bi-	two	binocular
bio-	life	biodiversity
cardi-	heart	cardiology
carni-	meat	carnivore
chloro-	green	chloroplast
di-	two	dioxide
eco-	environment	ecosystem
ex-	out of	exhale
geo-	earth	geologist
hemi-	half	hemisphere
hydro-	water	hydrogen
ign-	fire	igneous
in-	not	insoluble
infra-	underneath	infrared
kilo-	thousand	kilogram
meta-	change	metamorphosis
micro-	small	microorganism
sub-	under	submarine
terra-	land	terrarium
uni-	one	unicellular

Suffix	Meaning	Example
-al	relating to	tropical
-itis	disease or swelling	bronchitis
-meter	measure	thermometer
-ology	study	geology
-ped	foot	biped
-saur	lizard	dinosaur
-scope	instrument for viewing	microscope

Root	Meaning	Example
domin	rule	dominant
derm	skin	dermatologist
volv	turn	revolve
therm	heat	thermometer
luna	moon	lunar
nat	born	natural
rupt	break	eruption